Communications
in Computer and Information Science 1021

Commenced Publication in 2007
Founding and Former Series Editors:
Phoebe Chen, Alfredo Cuzzocrea, Xiaoyong Du, Orhun Kara, Ting Liu,
Krishna M. Sivalingam, Dominik Ślęzak, Takashi Washio, and Xiaokang Yang

Editorial Board Members

Simone Diniz Junqueira Barbosa🆔
  Pontifical Catholic University of Rio de Janeiro (PUC-Rio),
  Rio de Janeiro, Brazil
Joaquim Filipe🆔
  Polytechnic Institute of Setúbal, Setúbal, Portugal
Ashish Ghosh
  Indian Statistical Institute, Kolkata, India
Igor Kotenko🆔
  St. Petersburg Institute for Informatics and Automation of the Russian
  Academy of Sciences, St. Petersburg, Russia
Junsong Yuan
  University at Buffalo, The State University of New York, Buffalo, NY, USA
Lizhu Zhou
  Tsinghua University, Beijing, China

More information about this series at http://www.springer.com/series/7899

Martin Atzmueller · Wouter Duivesteijn (Eds.)

Artificial Intelligence

30th Benelux Conference, BNAIC 2018
's-Hertogenbosch, The Netherlands, November 8–9, 2018
Revised Selected Papers

 Springer

Editors
Martin Atzmueller
Tilburg University
Tilburg, The Netherlands

Wouter Duivesteijn
Eindhoven University of Technology
Eindhoven, The Netherlands

ISSN 1865-0929   ISSN 1865-0937 (electronic)
Communications in Computer and Information Science
ISBN 978-3-030-31977-9   ISBN 978-3-030-31978-6 (eBook)
https://doi.org/10.1007/978-3-030-31978-6

This Springer imprint is published by the registered company Springer Nature Switzerland AG
The registered company address is: Gewerbestrasse 11, 6330 Cham, Switzerland

Preface

This volume contains the proceedings of BNAIC 2018: the 30th Benelux Conference on Artificial Intelligence held in 's-Hertogenbosch, The Netherlands, during November 8–9, 2018. BNAIC was organized by the Jheronimus Academy of Data Science (JADS), under the auspices of the Benelux Association for Artificial Intelligence (BNVKI) and the Dutch Research School for Information and Knowledge Systems (SIKS).

For the proceedings, we selected high-quality papers from BNAIC itself, but also papers from the collocated Benelearn conference, based on quality and thematic fit. From the BNAIC submissions, we selected the Type A regular papers that had been accepted for oral presentation at BNAIC, which were also extended/revised for the proceedings. Each submission was reviewed by at least three Program Committee members. For the proceedings, we were able to include nine high-quality submissions from BNAIC. Further, we were able to include two high-quality contributions from the collocated Benelearn conference.

For the BNAIC 2018 proceedings, we would like to reiterate the credits and extend our thanks to many who supported the organization. The BNAIC 2018 conference would not have been possible without the support and efforts of many. We thank the members of the Program Committee for their constructive and scholarly reviews. We are grateful to Arjan van den Born, Chris Emmery, Arjan Haring, and Laura Niemeijer for their reliable administrative support and local organization/coordination at JADS. We also wish to thank all student volunteers for enthusiastically helping out in many ways. A special thanks goes to Eric Postma for his invaluable support for BNAIC 2018 at JADS in many ways behind the scenes.

Finally, we are grateful to our sponsors for their generous support of the conference:

- Target Holding
- DIKW Intelligence
- SNN Adaptive Intelligence
- SIKS
- BNVKI
- SKBS

July 2019

Martin Atzmueller
Wouter Duivesteijn

Organization

Program Committee

Stylianos Asteriadis	University of Maastricht, The Netherlands
Martin Atzmueller	Tilburg University, The Netherlands
Reyhan Aydogan	Delft University of Technology, The Netherlands
Floris Bex	Utrecht University, The Netherlands
Albert Bifet	LTCI, Telecom ParisTech, France
Tibor Bosse	Vrije Universiteit Amsterdam, The Netherlands
Bert Bredeweg	University of Amsterdam, The Netherlands
Tom Claassen	Radboud University, The Netherlands
Walter Daelemans	University of Antwerp, The Netherlands
Gregoire Danoy	University of Luxembourg, Luxembourg
Mehdi Dastani	Utrecht University, The Netherlands
Jesse Davis	Katholieke Universiteit Leuven, Belgium
Victor de Boer	Vrije Universiteit Amsterdam, The Netherlands
Marc Denecker	Katholieke Universiteit Leuven, Belgium
Wouter Duivesteijn	Eindhoven University of Technology, The Netherlands
Ad Feelders	Utrecht University, The Netherlands
George H. L. Fletcher	Eindhoven University of Technology, The Netherlands
Pascal Gribomont	University of Liege, Belgium
Perry Groot	Radboud University, The Netherlands
Andrew Hendrickson	Tilburg University, The Netherlands
Tom Heskes	Radboud University, The Netherlands
Arjen Hommersom	Open University of the Netherlands, The Netherlands
Mark Hoogendoorn	Vrije Universiteit Amsterdam, The Netherlands
Geert-Jan Houben	Delft University of Technology, The Netherlands
Kristian Kersting	TU Darmstadt, Germany
Arno Knobbe	Universiteit Leiden, The Netherlands
Walter Kosters	LIACS, Leiden University, The Netherlands
Johan Kwisthout	Radboud University, The Netherlands
Tom Lenaerts	Universite Libre de Bruxelles, Belgium
Peter Lucas	Leiden University, The Netherlands
Bernd Ludwig	University of Regensburg, Germany
Bernard Manderick	COMO Lab. Vrije Universiteit Brussel, Belgium
Elena Marchiori	Radboud University, The Netherlands
Wannes Meert	Katholieke Universiteit Leuven, Belgium
John-Jules Meyer	Utrecht University, The Netherlands
Aske Plaat	Leiden University, The Netherlands
Eric Postma	TiCC, Tilburg University, The Netherlands
Marie Postma	Tilburg University, The Netherlands

Henry Prakken	University of Utrecht and University of Groningen, The Netherlands
Stefan Schlobach	Vrije Universiteit Amsterdam, The Netherlands
Evgueni Smirnov	MICC-IKAT, Maastricht University, The Netherlands
Gerasimos Spanakis	Maastricht University, The Netherlands
Jennifer Spenader	University of Groningen, AI, The Netherlands
Johan Suykens	Katholieke Universiteit Leuven, Belgium
Annette Ten Teije	Vrije Universiteit Amsterdam, The Netherlands
Dirk Thierens	Utrecht University, The Netherlands
Jos Uiterwijk	Maastricht University, The Netherlands
Egon L. van den Broek	Utrecht University, The Netherlands
Jaap van Den Herik	Leiden University, The Netherlands
Peter van der Putten	LIACS, Leiden University and Pegasystems, The Netherlands
Leon van der Torre	University of Luxembourg, Luxembourg
Frank Van Harmelen	Vrije Universiteit Amsterdam, The Netherlands
M. Birna van Riemsdijk	Delft University of Technology, The Netherlands
Marieke van Vugt	University of Groningen, The Netherlands
Menno Van Zaanen	Tilburg University, The Netherlands
Remco Veltkamp	Utrecht University, The Netherlands
Joost Vennekens	Katholieke Universiteit Leuven, Belgium
Arnoud Visser	University of Amsterdam, The Netherlands
Willem Waegeman	Ghent University, Belgium
Gerhard Weiss	University Maastricht, The Netherlands
Marco Wiering	University of Groningen, The Netherlands
Jef Wijsen	University of Mons, Belgium
Mark H. M. Winands	Maastricht University, The Netherlands
Marcel Worring	University of Amsterdam, The Netherlands
Yingqian Zhang	Eindhoven University of Technology, The Netherlands

Additional Reviewers

Lapauw, Ruben
Menger, Vincent
Schraagen, Marijn
van der Hallen, Matthias

Contents

Early Detection of Sepsis Induced Deterioration Using Machine Learning

Francesco Dal Canton[1], Vincent M. Quinten[1,2], and Marco A. Wiering[1(✉)]

[1] University of Groningen, 9700 AB Groningen, The Netherlands
m.a.wiering@rug.nl
[2] University Medical Center Groningen, 9713 GZ Groningen, The Netherlands

Abstract. Sepsis is an excessive bodily reaction to an infection in the bloodstream, which causes one in five patients to deteriorate within two days after admission to the hospital. Until now, no clear tool for early detection of sepsis induced deterioration has been found. This research uses electrocardiograph (ECG), respiratory rate, and blood oxygen saturation continuous bio-signals collected from 132 patients from the University Medical Center of Groningen during the first 48 h after hospital admission. This data is examined under a range of feature extraction strategies and Machine Learning techniques as an exploratory framework to find the most promising methods for early detection of sepsis induced deterioration. The analysis includes the use of Gradient Boosting Machines, Random Forests, Linear Support Vector Machines, Multi-Layer Perceptrons, Naive Bayes Classifiers, and k-Nearest Neighbors classifiers. The most promising results were obtained using Linear Support Vector Machines trained on features extracted from single heart beats using the wavelet transform and autoregressive modelling, where the classification occurred as a majority vote of the heart beats over multiple long ECG segments.

Keywords: Sepsis · Machine Learning · Bio-signals · Health care

1 Introduction

Sepsis is a life-threatening organ dysfunction caused by an uncontrolled reaction to infection by the organism [1] that can lead to organ failure, septic shock, and death [2]. Common symptoms of sepsis include higher heart rate and respiratory rate, and abnormal changes in bodily temperature [3]. Sepsis is one of the most common causes for mortality among chronically ill patients, and it is estimated that sepsis affects at least 240 people out of 100,000 in the United States, while severe sepsis affects between 51 and 95 out of 100,000 [4]. Most patients affected by sepsis are admitted to the hospital through the Emergency Department (ED), and it was shown that approximately 20% of patients admitted to the ED with infection or sepsis deteriorate [5].

Early detection of sepsis induced deterioration is extremely valuable since it allows for fast and effective treatment. In [6] it was shown that each hour of delay

M. Atzmueller and W. Duivesteijn (Eds.): BNAIC 2018, CCIS 1021, pp. 1–15, 2019.
https://doi.org/10.1007/978-3-030-31978-6_1

in the application of appropriate treatment is correlated with a mean increase in mortality of 7.6%. Nevertheless, despite the intensive research in the field, it is still not clear how the onset, progress, and response to treatment of sepsis can be accurately monitored [7].

The traditional approach for tracking sepsis onset and development is to use discrete values describing vital signs and non-specific symptoms [3]. More recently, measures obtained from Heart Rate Variability (HRV) have been gathering research interest. Although at present the most successful studies in this area concerned sepsis development in neonates [8], some studies have been carried out to explore the predictive potential of HRV measures in adults [9,10]. In 2017 the SepsiVit study was started at the University Medical Center of Groningen (UMCG), which involves a long term data collection program, and aims at determining whether HRV measures can provide a reliable source of information for predicting deterioration in patients with suspected sepsis in the ED [11].

The current study focuses on the potential of Machine Learning based algorithms paired with the use of raw Electrocardiograph (ECG), Plethysmograph, and Respiratory Rate bio-signals collected during the SepsiVit study at the UMCG as sources of information for early detection of patient deterioration due to sepsis. Seven different Machine Learning classifiers are tested and their classification accuracies are compared across three different feature extraction methods. The first two methods involve Histograms of Derivatives (HOD) of the bio-signals, while the third one uses morphological features of heart beats extracted using the wavelet transform and autoregressive modelling as applied in [12]. The third feature extraction method was also tested in a majority vote fashion across 5 min long signal windows and 1 h long signal windows.

This paper is organized as follows. Section 2 describes the dataset in more detail. Section 3 illustrates the three feature extraction methods used to process the dataset. Section 4 lists and explains the machine learning models and how they were applied. Section 5 describes the experimental setup and the obtained results, while Sect. 6 concludes the paper.

2   Dataset

The dataset used in this research was collected at the ED of the UMCG according to the protocol of the SepsiVit study. All patients included in the study (i) are more than 18 years old, (ii) present a suspected infection or sepsis, (iii) show two or more systemic inflammatory response syndrome criteria as defined by the International Sepsis Definitions Conference [13], and (iv) provided written informed consent. Patients are not included in the study in case of (i) known pregnancy, (ii) when the patient is not admitted to the hospital from the ED or is transfered to another hospital or care facility, and (iii) in case of previous cardiac transplantation [11]. While the aim of the SepsiVit study is to collect data from 171 patients, the collected and labeled data at the time of the current study includes 132 patients (84 males; average age 61.5 years; median age 63.5 years; average missing data 53%).

For each patient, high sample rate vital signs are recorded with a bedside patient monitor (Philips IntelliVue MP70 System with MultiMeasurement Module using custom software based on the Philips IntelliVue Data Export Interface Protocol). The data includes time series data of ECG (500 Hz), Plethysmograph (125 Hz), and Respiratory Rate (62.5 Hz) bio-signals recorded for up to 48 h since admission to the ED. No imputation strategy is used to recover missing data due to the complexity and unpredictability of the bio-signals involved. The electrodes for recording the ECG signals are placed according to the EASI configuration [14], and in particular the data from Lead II is used for this analysis. After the data is collected, the outcomes for the patient's condition are recorded. Specifically, five outcomes are monitored: whether the patient (i) had to be transferred to the Intensive Care Unit (ICU), (ii) died in the hospital, (iii) developed kidney failure, (iv) developed liver failure, or (v) developed respiratory failure. Since the goal of this analysis is to provide a tool for early sepsis deterioration, each patient was labeled as 'deteriorating' if they registered positive to any of these five outcomes, and 'healthy' otherwise. The proportions of the two groups are specific to each feature extraction method depending on the amount of usable data, and are mentioned in the respective subsections of the paper.

3   Feature Extraction Methods

The detection of early signs of sepsis induced deterioration using bio-signals requires a procedure of feature extraction from the raw data, so that each extracted feature vector represents a segment of the original data. With this in mind, a good feature extraction procedure should yield feature vectors that are most similar among the same class and most different across different classes.

The three feature extraction methods described in this section are compared with the ones currently being developed as a part of the SepsiVit study, which were obtained exclusively from the ECG signal, after the removal of technical and physiological artifacts [15]. They include HRV measures as described in [16], and geometrical features of the R-R intervals [17].

3.1   Histograms of Derivatives

The first approach involves the extraction of the distribution of the first and second order derivatives of the available signals, or Histograms of Derivatives (HOD). This method is conceptually close to the Histogram of Oriented Gradients strategy used in image processing [18]: the objective is to obtain the frequency distribution of change in signal intensity across a signal segment. The derivative of a function at a specific input value is defined as the slope of the tangent line to the graph of the function at that point. In the case of the digital signals used in this study, an approximation of the derivative function is computed as:

$$\frac{dx}{dt} = \frac{x_{t+h} - x_t}{h} \tag{1}$$

where h is the unit interval between consecutive samples. For each of the three signals used in this study, h is set to 1 since the time between consecutive samples in each signal is constant.

The first step of this procedure is, for each patient's bio-signals (i.e. ECG, Plethysmograph, and Respiratory Rate), to extract all simultaneous 5-minute long signal segments that don't contain any missing data. The result is a collection of 5-minute long data triplets containing the three bio-signals. The length of 5 min for each signal window was chosen experimentally as it produced improved classification accuracies compared to a length of 30 min. This choice was also guided by the convenience of requiring only 5 min of recorded signal before attempting detection of sepsis induced deterioration, which would speed up the potential application of treatment.

At this stage, the first and second derivatives of each signal segment are computed. Given each signal in each data triplet, Eq. 1 was applied across the whole signal segment. The result is 6 signals, two for each type of bio-signal, of which one is the first order derivative, and the other is the second order derivative, computed by applying Eq. 1 on the computed first derivative. A plot representing an example of first and second order derivatives computed in such fashion is shown in Fig. 1.

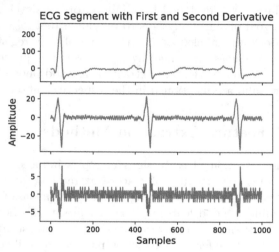

Fig. 1. Plot showing first and second order derivatives of an ECG signal segment taken from the SepsiVit dataset.

In order to obtain the frequency distribution of each derivative, a 20-bin frequency histogram is computed for each of the 6 derivative signals. In order to exclude outliers, the extrema of each histogram are computed as follows. For each of the 6 derivative signals, the minimum and maximum values are collected across the whole dataset, for a total of 12 values. A 95% interval is then calculated for each of the 12 resulting lists of values. The lowest value in the 95% interval was chosen for the minimum of each histogram, while the maximum value in the 95% interval was chosen for the maximum of each histogram. The values found with this method are reported in Table 1.

The result was six 20-bin histograms, three for the first derivative of ECG, Plethysmograph, and Respiratory Rate, and three for their second derivatives, for each 5-minute long data segment. Each of these histograms was then centered (by subtracting the mean) and scaled (by dividing by the standard deviation).

These six histograms were then concatenated so that the first three vectors were the histograms of the first derivative of ECG, Plethysmograph, and Respiratory Rate histograms, while the last three were the histograms of the second derivatives in the same order.

The last step of the feature extraction process involved, for the ECG signal contained in each of the data triplets, extracting the mean and the standard deviation of the Heart Rate, $\mu(HR)$ and $\sigma(HR)$. These two values were appended to each concatenated frequency histogram vector to produce a 122-dimensional feature vector. Only patients that had at least one uninterrupted 5-minute long window containing all three bio-signals were included in this procedure. This feature extraction method yielded 14,389 feature vectors from 89 different patients. Out of the total number of data triplets, 50.8% came from patients marked as 'deteriorating'.

3.2 Δ of Histograms of Derivatives

The second feature extraction approach is largely based on the one described in Subsect. 3.1. The objective of this method is to obtain a measure of the change between the HODs of consecutive 5-minute long data triplets. Initially all pairs of consecutive 5-minute long data triplets are collected, so that in each pair the second triplet directly follows the first one in the time domain. The two 122-dimensional feature vectors for both data triplets are then extracted according to the procedure described in Subsect. 3.1. The final feature vector is then computed as the element-wise difference between the two vectors as:

$$fv_\Delta = fv_t - fv_{t-1} \qquad (2)$$

where fv_{t-1} and fv_t are the feature vectors extracted from the first and second data triplets respectively. Only patients that had at least one uninterrupted 10-minute long window containing all three bio-signals were included in this procedure. This feature extraction procedure yielded 13,110 feature vectors from 88 different patients. Out of the total number of data triplets, 50.5% came from patients marked as 'deteriorating'.

3.3 Wavelet Transform and Autoregressive Modelling

The last feature extraction procedure involves using the wavelet transform and autoregressive modelling on exclusively the ECG signal. This approach relies on extracting morphological features from individual heart beats, replicating the approach found in [12]. This procedure required a preprocessing step of noise removal from the ECG signal and extraction of all available heart beats (done with the Python package Biosppy 0.5.1), where the R-peaks were detected using Hamilton's approach [19]. Each heart beat is extracted in the form of an array of 300 samples, where the R-peak occurs at the 100th sample. An example of a series of extracted heart beats is shown in Fig. 2.

Table 1. Extrema of each of the 6 frequency histograms, computed for the SepsiVit dataset by considering the 95% interval for each minimum and maximum value in each derivative signal.

	1st derivative		2nd derivative	
	Min	Max	Min	Max
ECG	−348	343	−307	307
Pleth.	−756	768	−511	518
Resp.	−681	722	−523	676

Due to memory limitations of the computer used when running the Machine Learning algorithms, a sample of 10,000 heart beats was selected for each patient to be used in the study. The sample of heart beats for each patient was selected by (1) extracting all heart beats for that patient, and (2) keeping 10,000 evenly spaced heart beats across all heart beats of the patient ordered in the time domain. This was done to ensure that, for each patient, heart beats from all stages of their stay in the hospital were available.

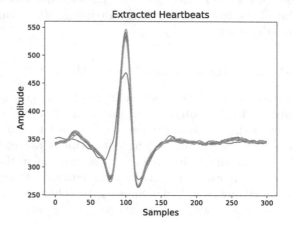

Fig. 2. Plot showing exemplar heart beats extracted from an ECG segment taken from the SepsiVit dataset, after noise removal has been applied. The different colors represent the different heart beats. (Color figure online)

A time-frequency decomposition of each heart beat was then produced using the wavelet transform as done in [12], which has been shown to be a good tool for QRS complex detection [20].

The wavelet transform is an operation that represents a signal with a series of coefficients which describe the energy distribution of the signal across both time and frequency. The continuous wavelet transform (CWT) of a continuous signal is defined as [21]:

$$CWT_x(b, a) = \frac{1}{\sqrt{|a|}} \int_{-\infty}^{\infty} x(t) g\left(\frac{t-b}{a}\right) dt \qquad (3)$$

where the wavelet $g(t)$ satisfies the conditions reported in [22]. a and b ($a, b \in \Re$, $a \neq 0$) are the dilation and translation parameters. The chosen wavelet, which in the case of this study is the Daubechies wavelet of order 8, as done by Qibin and Liqing [12], is compressed or expanded depending on the value of a, in such a way that coefficients can be extracted to describe the morphology of the signal at different frequency ranges. The high computational complexity of this approach can be reduced by discretising one or both parameters of the function. The case where a is discretised is defined as the dyadic wavelet transform D_yWT. a is discretised along the dyadic sequence 2^i ($i \in \mathbb{N}$) [20]. D_yWT is then defined as:

$$D_yWT_x(b, 2^i) = \frac{1}{\sqrt{2^i}} \int_{-\infty}^{\infty} x(t)g\left(\frac{t-b}{2^i}\right) dt \tag{4}$$

The dyadic wavelet transform was consequently applied to all heart beat signals (done with the Python package pywt 1.0.6 [23]). A required parameter for the operation was the decomposition level, which influences the frequency ranges extracted from the signal. The chosen decomposition level was 4 as done in [12]. The wavelet transform decomposition yielded four detail coefficients d_1, d_2, d_3, d_4 and the vector of approximation coefficients a_4. The detail coefficients represent the high frequency parts of the ECG signal, while the vector of approximation coefficients a_4 represent the lower frequency changes in each heart beat, corresponding with the main features of the QRS complexes. For each heart beat, the vector a_4 contained 32 points.

The second step was the extraction of the coefficients of an autoregressive model trained on each heart beat. An autoregressive model of order p of a signal $x[n]$ is defined as the linear combination of the p previous samples in the signal, and can be expressed as:

$$x[n] = \sum_{i=1}^{p} a[i]x[n-i] + e[n] \tag{5}$$

where $a[i]$ is the ith coefficient and $e[n]$ is white noise with mean zero [12]. The number of coefficients p was chosen to be 14 using the Akaike Information criterion [24], so that the 14 coefficients a_{ar} of the autoregressive model were extracted from each heart beat (done with the Python package statsmodels 0.9). The two obtained vectors $a_4 = \{w_1, \ldots, w_{32}\}$ and $a_{ar} = \{a_1, \ldots, a_{14}\}$ were then concatenated to form the feature vector for that heart beat. Only patients whose ECG signal contained at least one heart beat detectable using Hamilton's approach [19] were included in this procedure. This feature extraction procedure yielded 1,155,997 feature vectors from 123 different patients. Out of the total number of data triplets, 44.9% came from patients marked as 'deteriorating'.

Due to the large number of feature vectors obtained with this method, Principal Component Analysis (PCA), a common feature reduction procedure, was used to compress the dimensionality of the feature vectors from 46 to 10 dimensions [25]. PCA involves projecting a set of vectors across the dimension with the maximal variance, in order to reduce the number of dimensions while preserving the maximal amount of information regarding the distribution of the vectors. For each test, PCA was applied by fitting it on the training split of the data, and then applying it to both the training and the testing splits of the data.

4   Machine Learning Methods

All algorithms described in this section were implemented in Python using the package scikit-learn 0.19.1 [26]. The dataset was split into training and testing/validation sets using 90% and 10% of the data respectively. The strategy used for splitting the dataset was group 10-fold cross-validation, so that 10 iterations of testing were performed for each algorithm. An important property of the group k-folds strategy for dataset splitting is that no data from the same patient occurred in different folds, so as to eliminate overfitting over single patients. The results as reported in Sect. 5 consist of the mean classification accuracy for the tuned models across the 10 training iterations, along with its standard deviation. The accuracy was computed as the number of correct classifications over all classification attempts. For the Linear Support Vector Machine, weighted k-Nearest Neighbors, and Multi-Layer Perceptron, the data must be scaled. A MinMax scaler, which scales each feature to an interval $[0, 1]$, was chosen experimentally as it yielded better results compared to a standard scaler. For each training fold the scaler was fitted on the training split of the dataset, and consequently applied to both the training and the testing split. Class scaling was applied to the two classes in the training phase for all classifiers except for the Multi-Layer Perceptron and the Weighted k-Nearest Neighbors, in order to normalise the impact

Table 2. Parameters used for each of the classifiers. The feature extraction methods are, in order: Histograms of Derivatives (HOD, see Subsect. 3.1), Difference of Histograms of Derivatives (HOD$_\Delta$, see Subsect. 3.2), wavelet transform and autoregressive modelling (HB, see Subsect. 3.3), and using the HRV measures extracted as part of the SepsiVit study (SV). The classifiers are, in order: Linear Support Vector Machine (SVM), Random Forest (RF), Gradient Boosting Machine (GBM), Weighted k-Nearest Neighbors (WkNN), Multi-Layer Perceptron (MLP), and Linear Regression (LR).

		HOD	HOD$_\Delta$	HB	SV
SVM	C	11	12	15	9.5
RF	n_estimators	7,000	5,000	3,500	5,000
GBM	n_estimators	10,000	10,000	10,000	10,000
	learning_rate	0.01	0.01	0.005	0.0001
	min_samples.	10			
WkNN	n_neighbors	6	11	251	55
	p	1			
MLP	hidden_n.	31	53	7	4
	learning_rate	0.0005	0.0005	0.0005	0.001
	max_iter	3,000			
	activation	*logistic*			
LR	C	15	8	10	15
	solver	*newton-cg*			
	multi_class	*multinomial*			

of the distribution of the two classes during training. The parameter tuning for all algorithms was done by parameter grid search using cross-validation. The parameters for all algorithms are reported in Table 2.

4.1   Linear Support Vector Machine

Support Vector Machines (SVMs) are a set of supervised learning algorithms useful in classification, which is widely and successfully applied in the medical field [12,27,28]. A Linear Support Vector Machine generates a hyperplane which position and orientation is optimised to best differentiate between the two classes, and which is computed using the support vectors, which are the vectors in the training set closest to the decision hyperplane [29]. The Linear SVM model used the squared hinge loss function, which produced a classification boundary with a soft margin, yielding classification probabilities. The only tuned parameter was C, which represents the importance given to outliers during training.

4.2   Random Forest

A Random Forest is an ensemble-based algorithm which works as a combination of decision tree predictors [30]. Each tree in a Random Forest is initialised using the values of a random vector sampled independently using the same distribution. This method is more robust to overfitting compared to standard decision trees [31]. All default parameters were kept the same as the scikit-learn implementation of the algorithm [26], except for $n\_estimators$, the number of trees to be generated. As the number of trees is increased, the accuracy normally increases and eventually plateaus. In the case of the wavelet transform and autoregressive modelling feature extraction method (see Subsect. 3.3), the number of generated trees was artificially kept low to accomodate for the memory limitations of the computer used in the analysis.

4.3   Gradient Boosting Machine

The Gradient Boosting Machine algorithm is, much like the Random Forest, an ensemble-based algorithm used in classification which combines a number of weak decision tree classifiers into a strong decision tree classifier. Each decision tree is generated by combining the previous decision trees and applying a higher weight to events that are difficult to predict. The result is a gradient descent algorithm that minimizes the classification error by generating more decision trees [32]. The two parameters that were tuned for this algorithm were $n\_estimators$, the number of trees to be generated, and $learning\_rate$, which shrinks the contribution of each tree. There is a trade-off between the values of the two parameters, so they need to be adjusted to each other. For all other parameters, the defaults of the scikit-learn package were used, except for the value of $min\_samples\_leaf$, which was set to 10. This value defines the minimum number of feature vectors to be found in each leaf of the decision trees.

4.4   Weighted k-Nearest Neighbors

The Weighted k-Nearest Neighbors (WkNN) algorithm is a variation of the standard k-Nearest Neighbors classification algorithm. The latter works by, for each feature vector in the testing set, producing a majority vote across the k closest feature vectors of the training set, according to a specified distance metric. The WkNN algorithm works in a similar fashion, with the added feature that votes from each neighboring feature vector are scaled depending on their distance from the feature vector to be classified [33]. The tuned parameter was only $n\_neighbors$, which is k, the number of the closest feature vectors that are taken into account for the classification. The distance metric used for this algorithm was the Minkowski distance, with the inverse scaling factor p set to 1.

4.5   Multi-Layer Perceptron

The Multi-Layer Perceptron (MLP) is a type of feedforward artificial neural network which implements the backpropagation supervised learning algorithm. The MLP implemented as a part of this study contained only one hidden layer. The amount of neurons in the hidden layer was the parameter $hidden\_neurons$, tuned for each feature extraction method. The final, output layer contains a number of neurons equal to the number of classes, to which activations a Softmax function is applied in order to compute class-wise probabilities. The $learning\_rate$ parameter was also tuned using cross-validation [31,34]. All other parameters were kept to the defaults given by scikit-learn, except for the applied logistic activation function, and the maximum number of training iterations for the algorithm, which was set to 3,000.

4.6   Naïve Bayes Classifier

The Naïve Bayes classifier is one of the simplest probabilistic classifiers, which has the advantage of being computationally inexpensive, and has been used with success on Heart Rate Arrhythmia classification in [35]. This classifier constructs a set of probabilities, which correspond to the probability that each feature value appears among the feature vectors within a certain class. The Naïve Bayes classifier makes, however, a strong assumption of conditional independence between the features within the feature vectors [36]. This assumption rarely holds in real life scenarios, and it clearly doesn't hold for the feature vectors extracted with the procedures described in Sect. 3. For this study, the Gaussian Naïve Bayes classifier was used, which relies on the assumption that the likelihood of the features follows a Gaussian distribution. The algorithm was tested as it tends to perform well in many classification tasks, and because of its conveniently low computational complexity. This classifier requires only the prior probabilities of the two classes, computed as the proportion of each class across each complete processed dataset.

4.7   Logistic Regression

The Logistic Regression classifier is a standard linear model for classification. In this study, a multinomial logistic regression was used, which means that the probability estimates should be better calibrated per class compared to a dichotomous implementation. The classifier used the 'newton-cg' solver. The only parameter tuned using cross-validation was C, the inverse of the regularization strength α.

5   Experiments and Results

For each tuned classifier and for every testing procedure, the mean and standard deviation of the classification accuracy across the 10 folds of the cross-validation process are reported. The testing procedures were five in total. The first three involved standard classification of the feature vectors obtained with the three feature extraction methods described in Sect. 3 using cross-validation. For each of the three produced datasets, each feature vector was assigned the same label as the patient that it was extracted from. During the training phase, the classifier was trained on the training set using the correct labels. During the testing phase, each feature vector was classified as belonging to the 'deteriorating' class or to the 'healthy' class. The result of the classification was then compared with the correct label in order to compute the accuracy (i.e. the proportion of correct classifications during the testing phase).

The last two testing procedures were applied to the morphology descriptors, which are described in Subsect. 3.3). For both testing procedures, the training phase was the same as for the third testing procedure, so that the classifier could classify each heart beat as 'deteriorating' or not given its feature vector. What changed in the last two testing procedures was the testing phase. The first of the two testing procedures was done as a majority vote, where heart beats are extracted and processed for all 5-minute long ECG segments. The classification process is then applied to all heart beats in each 5-minute long ECG segment so that if 50% or more of the heart beats are classified as 'deteriorating', then the whole segment receives such classification outcome. The third testing procedure is performed in a similar fashion by taking a majority vote across 12 5-minute long ECG segments.

All testing procedures are compared to the performance of the tuned algorithms used on the HRV features extracted as part of the SepsiVit study, as mentioned in Sect. 3. All outcomes of the testing procedures are reported in Table 3.

The Histograms of Derivatives and Differences of Histograms of Derivatives methods for feature extraction did not show any promise, ranging from a mean classification accuracy of $43.1 \pm 11.9\%$ for the Multi-Layer Perceptron in the Difference of Histograms of Derivatives procedure, to $56.6 \pm 12\%$ for the Random Forests algorithm applied to the Histograms of Derivative method for feature extraction.

The best results were obtained using the Linear Support Vector Machine on the feature vectors extracted in the SepsiVit study, which had a mean accuracy

Table 3. Mean and standard deviation of the classification accuracies for all models and testing procedures. The testing procedures are, in order: Histograms of Derivatives (HOD, see Subsect. 3.1), difference of Histograms of Derivatives (HOD$_\Delta$, see Subsect. 3.2), wavelet transform and autoregressive modelling without majority vote (HB, see Subsect. 3.3), wavelet transform and autoregressive modelling applied in a majority vote fashion over 5-minute long ECG segments (MV), wavelet transform and autoregressive modelling applied in a majority vote fashion over 12 5-minute long ECG segments (MV$_2$), and using the HRV measures extracted as part of the SepsiVit study (SV). The classifiers are, in order: Linear Regression (LR), Weighted k-Nearest Neighbors (WkNN), Naïve Bayes (NB), Linear Support Vector Machine (SVM), Multi-Layer Perceptron (MLP) Random Forest (RF), and Gradient Boosting Machine (GBM).

	HOD	HOD$_\Delta$	HB	MV	MV$_2$	SV
LR	54.1 ± 14.3	50.5 ± 7.4	59.3 ± 9.4	60.6 ± 10.8	61.0 ± 10.6	63.0 ± 5.2
WkNN	52.8 ± 6.7	50.4 ± 6.6	55.1 ± 6.3	57.1 ± 10.9	57.8 ± 11.2	57.9 ± 5.8
NB	54.8 ± 13.3	49.7 ± 14.3	51.8 ± 10.7	54.0 ± 15.4	53.9 ± 15.9	57.9 ± 5.8
SVM	52.4 ± 13.9	50.5 ± 12.7	$\mathbf{60.9 \pm 9.1}$	$\mathbf{62.2 \pm 10.7}$	$\mathbf{62.4 \pm 10.9}$	$\mathbf{65.5 \pm 7.9}$
MLP	53.8 ± 11.1	43.1 ± 11.9	59.8 ± 12.9	57.1 ± 15.2	56.9 ± 15.9	60.3 ± 8.1
RF	$\mathbf{56.3 \pm 12}$	$\mathbf{54.8 \pm 6.7}$	55.4 ± 7.8	58.2 ± 12.2	58.5 ± 12.8	59.3 ± 6.9
GBM	54.6 ± 8.4	54.4 ± 9.0	57.6 ± 7.8	61.5 ± 13.1	61.9 ± 13.6	61.3 ± 8.5

of 65.5% and a standard deviation of 7.9%. The most promising results were obtained with the feature extraction method involving the wavelet transform and autoregressive modelling, which was only marginally improved by the majority vote testing procedures. The Linear Support Vector Machine classifier produced the best results with the data extracted in this fashion, peaking at $62.4 \pm 10.9\%$ mean classification accuracy.

Overall, the Linear Support Vector Machine was the best classifier, sometimes beaten by the Random Forest.

6   Conclusion and Future Work

The results presented in the previous section show that none of the attempted feature extraction methods are superior in their ability to encapsulate differences between the two classes and similarity among the same class compared to the HRV features extracted as part of the SepsiVit study [11]. Nonetheless, the results of this study imply that there is more useful information in the morphological descriptions of the ECG signal compared to the frequency distributions of the slopes of high frequency bio-signals.

While there was an increase in classification accuracy obtained by applying the majority vote testing strategies, the fact that the improvement was as small as 1.5% indicates that the improvement is only marginal, and given the benefits of early detection of sepsis induced deterioration [6], a classification strategy requiring less data such as the standard heart beat classification or the majority vote across 5-minute ECG segments might be more beneficial for improving survival rates, compared to one that uses 60-minute ECG segments.

A difficulty encountered in this study was the limited size of the dataset. The low variability in the bio-signals across the data of each individual patient makes it so that the diversity in the dataset, and so the capacity of the Machine Learning algorithms to properly generalise the problem, is entirely dependent on the amount of different patients included in the study. Since reaching the target of the SepsiVit study of 171 patients (i.e. only 30% more than were available for this research) is likely not going to produce sufficient diversity in the dataset, future data collection programs are needed to further investigate the predictive potential of high frequency bio-signals for early detection of sepsis induced deterioration.

Future studies could focus on any of the following points for improvement. A more complete analysis of the feature extraction methods should be carried out: new strategies should be tested, and all strategies should be used together to produce feature vectors containing all features for each bio-signal segment. An analysis of which features contribute the most to the classification would then reveal the features that are most relevant towards the early detection of sepsis induced deterioration. Furthermore, different classifiers should be tested. Obvious candidates are Recurrent Neural Networks such as LSTMs, widely used on time series data, which nevertheless require large amounts of data for effective training, and which as such would depend on a new data collection program.

References

1. Singer, M., et al.: The third international consensus definitions for sepsis and septic shock (sepsis-3). JAMA **315**(8), 801–810 (2016). 26903338[pmid]
2. Bone, R.C., Fisher, C.J., Clemmer, T.P., Slotman, G.J., Metz, C.A., Balk, R.A.: Sepsis syndrome: a valid clinical entity. Methylprednisolone severe sepsis study group. Crit. Care Med. **17**(5), 389–393 (1989)
3. Buchan, C.A., Bravi, A., Seely, A.J.E.: Variability analysis and the diagnosis, management, and treatment of sepsis. Curr. Infect. Dis. Rep. **14**(5), 512–521 (2012)
4. Danai, P., Martin, G.S.: Epidemiology of sepsis: recent advances. Curr. Infect. Dis. Rep. **7**(5), 329–334 (2005)
5. Glickman, S.W., et al.: Disease progression in hemodynamically stable patients presenting to the emergency department with sepsis. Acad. Emerg. Med. **17**(4), 383–390 (2010)
6. Brindley, P.G., Zhu, N., Sligl, W.: Best evidence in critical care medicine early antibiotics and survival from septic shock: it's about time. Can. J. Anesth./Journal canadien d'anesthésie **53**(11), 1157–1160 (2006)
7. Dellinger, R.P., et al.: Surviving sepsis campaign: international guidelines for management of severe sepsis and septic shock 2012. Crit. Care Med. **41**(2), 580–637 (2013)
8. Moorman, J.R., et al.: Mortality reduction by heart rate characteristic monitoring in very low birth weight neonates: a randomized trial. J. Pediatrics **159**(6), 900–906.e1 (2011)
9. Ahmad, S., et al.: Continuous multi-parameter heart rate variability analysis heralds onset of sepsis in adults. PLoS ONE **4**(8), 1–10 (2009)

10. Bravi, A., Green, G., Longtin, A., Seely, A.J.E.: Monitoring and identification of sepsis development through a composite measure of heart rate variability. PLoS ONE **7**(9), e45666 (2012). PONE-D-12-18432[PII]
11. Quinten, V.M., van Meurs, M., Renes, M.H., Ligtenberg, J.J.M., ter Maaten, J.C.: Protocol of the SepsiVit study: a prospective observational study to determine whether continuous heart rate variability measurement during the first 48 hours of hospitalisation provides an early warning for deterioration in patients presenting with infec. BMJ Open **7**(11), e018259 (2017)
12. Zhao, Q., Zhang, L.: ECG feature extraction and classification using wavelet transform and support vector machines. In: 2005 International Conference on Neural Networks and Brain, vol. 2, pp. 1089–1092, October 2005
13. Levy, M.M., et al.: 2001 SCCM/ESICM/ACCP/ATS/SIS international sepsis definitions conference. Crit. Care Med. **31**(4), 1250–1256 (2003)
14. Cardoso, J.F., Laheld, B.H.: Equivariant adaptive source separation. IEEE Trans. Signal Process. **44**(12), 3017–3030 (1996)
15. Peltola, M.: Role of editing of R-R intervals in the analysis of heart rate variability. Front. Physiol. **3**, 148 (2012)
16. Shaffer, F., Ginsberg, J.P.: An overview of heart rate variability metrics and norms. Front. Public Health **5**, 258 (2017). 29034226[pmid]
17. Moridani, M.K., Setarehdan, S.K., Nasrabadi, A.M., Hajinasrollah, E.: Non-linear feature extraction from HRV signal for mortality prediction of ICU cardiovascular patient. J. Med. Eng. Technol. **40**(3), 87–98 (2016). PMID: 27028609
18. Dalal, N., Triggs, B.: Histograms of oriented gradients for human detection. In: 2005 IEEE Computer Society Conference on Computer Vision and Pattern Recognition (CVPR 2005), vol. 1, pp. 886–893, June 2005
19. Hamilton, P.: Open source ECG analysis. Comput. Cardiol. **29**, 101–104 (2002)
20. Kadambe, S., Murray, R., Boudreaux-Bartels, G.F.: Wavelet transform-based QRS complex detector. IEEE Trans. Biomed. Eng. **46**(7), 838–848 (1999)
21. Morlet, J., Arens, G., Fourgeau, E., Glard, D.: Wave propagation and sampling theory - Part i: complex signal and scattering in multilayered media. Geophysics **47**(2), 203–221 (1982)
22. Grossmann, A.: Wavelet transforms and edge detection. In: Albeverio, S., Blanchard, P., Hazewinkel, M., Streit, L. (eds.) Stochastic Processes in Physics and Engineering, pp. 149–157. Springer, Dordrecht (1988). https://doi.org/10.1007/978-94-009-2893-0_7
23. Lee, G., et al.: Pywavelets - wavelet transforms in Python (2006). Accessed 2018
24. Akaike, H.: Information theory and an extension of the maximum likelihood principle. In: Parzen, E., Tanabe, K., Kitagawa, G. (eds.) Selected Papers of Hirotugu Akaike, pp. 199–213. Springer, New York (1998). https://doi.org/10.1007/978-1-4612-1694-0_15
25. Jolliffe, I.: Principal component analysis. In: Lovric, M. (ed.) International Encyclopedia of Statistical Science, pp. 1094–1096. Springer, Heidelberg (2011). https://doi.org/10.1007/978-3-642-04898-2
26. Pedregosa, F., et al.: Scikit-learn: machine learning in Python. J. Mach. Learn. Res. **12**, 2825–2830 (2011)
27. Li, Q., Rajagopalan, C., Clifford, G.D.: Ventricular fibrillation and tachycardia classification using a machine learning approach. IEEE Trans. Biomed. Eng. **61**(6), 1607–1613 (2014)
28. Song, M.H., Lee, J., Cho, S.P., Lee, K.J., Yoo, S.K.: Support vector machine based arrhythmia classification using reduced features. Int. J. Control Autom. Syst. **3**(4), 571–579 (2005)

29. Hearst, M.A., Dumais, S.T., Osuna, E., Platt, J., Schölkopf, B.: Support vector machines. IEEE Intell. Syst. Appl. **13**(4), 18–28 (1998)
30. Breiman, L.: Random forests. Mach. Learn. **45**(1), 5–32 (2001)
31. Hastie, T., Tibshirani, R., Friedman, J.H.: The Elements of Statistical Learning. Springer, New York (2009). https://doi.org/10.1007/978-0-387-84858-7
32. Friedman, J.H.: Greedy function approximation: a gradient boosting machine. Ann. Stat. **29**(5), 1189–1232 (2001)
33. Hechenbichler, K., Schliep, K.: Weighted k-nearest-neighbor techniques and ordinal classification (2004). Accessed 2018
34. Kriesel, D.: A brief introduction to neural networks (2007)
35. Soman, T., Bobbie, P.O.: Classification of arrhythmia using machine learning techniques. WSEAS Trans. Comput. **4**, 548–552 (2005)
36. Chan, T.F., Golub, G.H., LeVeque, R.J.: Updating formulae and a pairwise algorithm for computing sample variances. In: Caussinus, H., Ettinger, P., Tomassone, R. (eds.) COMPSTAT 1982 5th Symposium Held at Toulouse 1982, pp. 30–41. Physica-Verlag, Heidelberg (1982)

Deriving Formulas for Integer Sequences Using Inductive Programming

Les De Ridder[✉][iD] and Thijs Vercammen[iD]

Department of Computer Science, KU Leuven, 3000 Leuven, Belgium
{les.deridder,thijs.vercammen}@student.kuleuven.be

Abstract. Solving integer sequences, correctly predicting the next number in a given sequence, is a challenging task for both humans and artificial intelligence. We present a method to derive a formula for an integer sequence given a subsequence. By splitting the known subsequence into 'windows', we can derive constraints in the form of linear combinations, which can be generalised to find a formula for the complete sequence. This approach is effective and can compete with existing methods based on pattern recognition and Artificial Neural Networks with regard to performance, success rate, and output quality.

Keywords: Integer sequences · Number series · Inductive programming · Linear combinations

1 Introduction

Predicting the next number in a given integer sequence, or *solving* a number series, is a challenging task for humans, often used in intelligence tests, and for Artificial Intelligence alike [2].

In this paper we present a method to derive a formula for an integer sequence when given a known subsequence, by finding appropriate linear combinations of the given numbers. We use partitions of the given subsequence as learning examples and induce a formula from the found linear combinations.

Existing solutions for this problem use e.g. functional programming with pattern recognition to derive a function or program for a sequence [3], or artificial neural networks to find the next number of a sequence [5].

A valuable resource on integer sequences is the 'Online Encyclopedia of Integer Sequences' (OEIS) [4]. It classifies over 300,000 sequences into various categories based on importance, difficulty of solving, mathematical properties, etc.

This paper is based on our bachelor's thesis, and is organised as follows: we first introduce the problem, next we outline our proposed solution, then we present an evaluation of our algorithm, and finally we compare our solution to existing methods.

© Springer Nature Switzerland AG 2019
M. Atzmueller and W. Duivesteijn (Eds.): BNAIC 2018, CCIS 1021, pp. 16–24, 2019.
https://doi.org/10.1007/978-3-030-31978-6_2

2   Problem Statement

The problem that we solve can be described as follows:
Take an integer sequence **a** with length n^1,

$$(a_1, a_2, \ldots, a_{n-1}, a_n),$$

given a subsequence of **a**, find a formula for all a_i, with $i = 1..n$.

This problem is an extension of the problem where given m consecutive elements of a sequence, starting with the kth element, the $(k + m)$th element is searched. By finding a formula for the whole subsequence instead, not only the next, $(k + m)$th element can be found, but also any other element.

Our algorithm will in principle only derive recursive formulas for a given input sequence, expressing a_i in terms of $a_{i-1}, \ldots, a_1$.

3   Approach

3.1   Windows

To approach this problem, we introduce the concept of windows on sequences. A window with length w 'selects' a subsequence of w consecutive integers of the input sequence. This window can be moved to create new selections of the input sequence. For every possible window size, we thus obtain a list of subsequences, generated by the sliding window.

3.2   Linear Combinations

After partitioning the input sequence by using a sliding window, we make the assumption that the last number of each window can be written as a linear combination of the remaining numbers of the window, with the same unknown coefficients for each window. This way, each a_i can be written as a linear combination of $a_{i-1}, \ldots, a_1$.

3.3   Feature Vectors

We can find formulas for sequences that can't be described by simple linear combinations of numbers in the window by adding extra *feature vectors*. More concretely, we take a list of functions as an extra parameter for the input of our algorithm.

These functions can be operations on numbers of the window, e.g. an exponentiation, the product of the numbers in the window (except the last one), or the function can be e.g. a constant function.

---

[1] Note that n can often be infinite.

We use the function results as extra terms in the linear combinations, for which new unknowns are introduced. For a sequence **a** consisting of numbers $a_1, a_2, \ldots, a_n$, this produces an equation of the form

$$
\begin{aligned}
a_i = &\ c_1 \cdot a_{i-1} + c_2 \cdot a_{i-2} + \cdots + c_{w-1} \cdot a_{i-w+1} \\
&+ d_1 \cdot f(a_{i-1}) + d_2 \cdot f(a_{i-2}) + \cdots + d_{w-1} \cdot f(a_{i-w+1}) \\
&+ e \cdot g(\{a_{i-1}, a_{i-2}, \ldots, a_{i-w+1}\}) \\
&+ q
\end{aligned}
$$

Here c_i, d_i and e are the (unknown) coefficients belonging to their respective terms, f is a function that applies to single numbers in the window, g is a function that applies to all the numbers in the window and q is a constant.

3.4   Algorithm

We start the algorithm with an initial window size of 2. We calculate the system of linear equations and solve it with a linear system solver. If we find a solution, we reconstruct the formula for the integer sequence from the found coefficient vector, and return the resulting formula. If the system was not solvable, we increase the window size by 1, we construct the new system and try again. If the maximally permissible window size is reached and we haven't found a solution yet, the algorithm fails.

A description of this algorithm in pseudocode is given in Algorithm 1.

3.5   Limitations

By using the appropriate functions as input for the algorithm, it becomes possible to find formulas for a significantly larger part of possible input sequences than without extra feature vectors.

The algorithm is however still limited since it mainly derives recursive formulas. There are integer sequences for which no trivial recursive formula exists or is known. Furthermore, the computational complexity is dependent on the calculations of the added functions, which can have a large impact on the run time of the algorithm (for certain input functions, or for large numbers of input functions).

Finally, using these extra feature vectors causes the number of columns in the coefficient matrix to rise. This means that to keep the system determined, the number of rows in the coefficient matrix has to be higher than in the case where no extra feature vectors are used. The input sequence therefore has to be longer when more feature vectors are used simultaneously. When there are insufficient rows, it's naturally also possible to choose a solution from the infinite solution set, but this is unlikely to be the solution that is sought.

Algorithm 1. Pseudocode of the algorithm presented in Section 3.4

```
 1:     ▷ In this pseudocode, we assume all input functions are functions that apply to
        all numbers in the window (for brevity).
 2:
 3:  procedure FINDFORMULA(InputSeq, InputFuns)
 4:      for WinSize ← 2 to Length(InputSeq) do
 5:          System ← GENERATELINEARSYSTEM(InputSeq, WinSize, InputFuns)
 6:          Solution ← Solve(System)
 7:          if Solution ≠ ∅ then
 8:              Derive Formula from Solution and InputFuns
 9:              return Formula
10:          end if
11:      end for
12:      return ∅
13:  end procedure
14:
15:  function GENERATELINEARSYSTEM(InputSeq, WinSize, InputFuns)
16:                          ▷ C_i are the unknown coefficients of our linear combinations.
17:      System ← ∅
18:      for i ← 1 to Length(InputSeq) − WinSize + 1 do
19:          Equation ← InputSeq_{i+WinSize−1} = C_1 InputSeq_i + C_2 InputSeq_{i+1} + ⋯ +
             C_{WinSize−2} InputSeq_{i+WinSize−3} + C_{WinSize−1} InputSeq_{i+WinSize−2}
20:          for f_j in InputFuns do
21:              Equation.RHS ← Equation.RHS + C_{f_j} f_j({InputSeq_{i..(i+WinSize−2)}})}
22:          end for
23:          System ← System ∪ {Equation}
24:      end for
25:      return System
26:  end function
```

4 Implementation

For a given input sequence **a** with length n, window size m, and function values
of the p feature vectors f, a linear system solver is given the coefficient matrix
and the vector of constant terms (last number of each window) as input:

$$A = \begin{bmatrix} a_1 & \cdots & a_{m-1} & f_{1,1} & \cdots & f_{p,1} \\ a_2 & \cdots & a_m & f_{1,2} & \cdots & f_{p,2} \\ \cdots & \cdots & \cdots & \cdots & \cdots & \cdots \\ a_{n-m+1} & \cdots & a_{n-1} & f_{1,n-m+1} & \cdots & f_{p,n-m+1} \end{bmatrix}, \quad b = \begin{bmatrix} a_m \\ a_{m+1} \\ \cdots \\ a_n \end{bmatrix}.$$

These are used by the solver to solve the matrix equation $Ax = b$. If a solution exists, we use the solution vector x to construct a formula for the integer sequence.

5 Experiments

We benchmarked the performance of our algorithm and implementation using sequences from the 'Online Encyclopedia of Integer Sequences' (OEIS) [4], a database with over 300,000 classified sequences. For performance reasons, we trimmed all sequences longer than 16 elements to the first 16 numbers.

For a sequence of n numbers, we run our algorithm on the first $n-1$ numbers and calculate the nth number using the formula found by our algorithm. We consider a found formula for a sequence to be correct if the calculated number is equal to the nth number in the input sequence.

Note that this doesn't necessarily mean that a formula found by our algorithm determines the same sequence as the sequence from the OEIS. For example, some sequences contain a repetition of the same number as long as or longer than the input subsequence size. In this case our implementation will derive a constant function for the sequence. We still consider these cases to be correct because our implementation finds a correct solution for the input subsequence.

To solve the problem of having too many sequences used at once, described in Sect. 3.5, we use three sets of feature vectors. When the algorithm does not succeed at finding a solution with the first set, the second and if necessary the third set are tried. The first set contains only the constant function. The second set contains a function that squares all the numbers in the window. The third set contains a function that returns the position of the last number of the window in the original input sequence, and the constant function.

5.1 Sequences in OEIS Categories

We test our implementation on sequences from the *core, easy, hard, nice, base* and *fini* categories of the OEIS. Table 1 shows an example sequence for each used category. For each of these categories we used a sample of 500 sequences, except for the *core* category where we used the full set of 174 sequences.

It's clear from the results (Table 2) that the ability of our algorithm to find a formula for a given sequence depends on the category of the sequence. As expected, our implementation was able to find a formula for only 3% of the sequences in the *hard* category, mostly sequences with a repetitive start. Our implementation works better on sequences from the *easy* category, where it was able to find a formula for almost a third of the sequences in the sample. In the *core* category, with the most important sequences, our implementation can find a formula for a quarter of the sequences.

Table 1. Examples of OEIS sequences by category

Category	Sequence	Description	OEIS identifier
core	1, 2, 3, 4, 5, 6, 7, ...	Natural numbers	A000027
easy	1, 2, 4, 8, 16, 32, 64, ...	Powers of 2	A000079
hard	2, 3, 5, 7, 13, 17, 19, ...	Mersenne exponents	A000043
nice	0, 1, 1, 2, 3, 5, 8, ...	Fibonacci sequence	A000045
base	0, 1, ..., 11, 22, ..., 101, ...	Palindromes in base 10	A002113
fini	1, 2, 3, 4, 6, 8, 12, 24	Divisors of 24	A018253

Table 2. Results of our algorithm on the OEIS

Category	Total	Tested	Solved	Percentage
core	174	174	47	27.01%
easy	74611	500	164	32.80%
hard	6319	500	16	3.20%
nice	6835	500	132	26.40%
base	35369	500	43	8.60%
fini	5977	500	28	5.60%

5.2 Input Length

We now investigate what the effect is of using shorter sequences on our algorithm's ability to solve them. We run our implementation on sequences of the *core* category, shortened to first 2, then 3, then 4, ... input numbers. The results are shown in Fig. 1.

When the input sequence has a length of 2 or 3, our algorithm can often find a correct third or fourth number because a lot of the sequences have a trivial start. Afterwards, when the sequences start to show complexity, the number of sequences for which our algorithm can find a formula drops quickly. This is because the window cannot be shifted enough for short sequences. When the input gets a bit longer, the window can be shifted more times and more formulas can be found again.

Furthermore the number of sequences for which we can successfully find a formula fluctuates mildly. This is a consequence of solutions that are correct for the next number in the sequence, but not for the numbers after it.

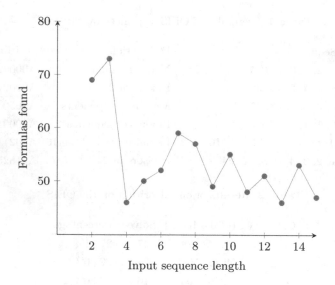

Fig. 1. Number of sequences for which a formula can be found per sequence length

5.3 Feature Vectors

Table 3 shows the number of sequences from the core category for which a formula can be found when only one particular feature vector gets used and when all three feature vectors get used at the same time.

Note that only 43 sequences are found when all feature vectors get used at the same time, compared to 47 in the previous results. This illustrates the problem with using too many feature vectors described in Sect. 3.5.

Interesting is that the algorithm can find multiple formulas for the same sequence, depending on the feature vectors used. A simple example of this is the sequence of natural numbers, $(1, 2, 3, 4, 5, 6, \ldots)$, which has the solution $\{a_n = -a_{n-2} + 2 \times a_{n-1}\}$ without feature vectors, but the solution $\{a_n = -a_{n-1} + 1\}$ with the constant function as a feature vector.

Table 3. Number of solved sequences from the *core* category of the OEIS per feature vector

Feature	Sequences solved
None	38
Constant	39
Square	41
Position in sequence	43
All	43

6   Comparison with Other Methods

6.1   MagicHaskeller

An earlier method [1] uses a functional method with an analytical generate and test approach, implemented in MagicHaskeller. A set of twenty self made sequences with a length of 5 to 7 numbers each was used to test the MagicHaskeller implementation. This implementation succeeded in finding a formula for 8 out of 20 sequences with an execution time varying from a few milliseconds to several dozens of seconds.

In the first test, our implementation succeeded in finding a formula for 13 sequences. Our algorithm had issues with 'alternating' sequences: sequences that alternate between two or more operations on the previous number. These sequences need a relatively large window to be described with a recursive formula and with a sequence length of 5 to 7 numbers, our algorithm can't move the window enough to generate the necessary equations. We extended these sequences to 10 numbers and this time, our implementation succeeded in finding a formula for 17 out of 20 sequences with a total execution time of 2.5 s.

6.2   IGOR

Another functional method [3], named IGOR, generates Haskell programs via pattern recognition. A self-made set of 20 sequences was used here as well, each with a length of 5 numbers. The IGOR implementation succeeded in generating a program for 13 sequences, with an execution time of maximum 1 h and 25 min per sequence.

Our implementation had once again a problem with alternating sequences and was only able to find a formula for 10 sequences at first. After extending the sequences, our implementation was able to find a formula for 14 out of 20 sequences with a total execution time of 2 s.

6.3   Neural Networks

Artificial Neural Networks (ANN) have also been used to solve integer sequences [5]. Both a set of 20 self-made sequences with a length of 5, as well as a subset of 57.000 integer sequences from the OEIS were used to benchmark the ANN implementation. Out of the set of 20, the neural network was able to predict the next number for 17 of the sequences.

Our implementation was able to find a formula for 16 of the sequences initially and 18 after extending the sequences to 10 input numbers.

Out of the set of 57.000 sequences, the best neural network could find the next number for almost 13.000 sequences. All generated neural networks together could predict the correct next number for almost 27.000 sequences. We cannot make a direct comparison because we don't know which sequences of the OEIS they used, but judging from the poor performance of our implementation in some categories, we can conclude that our implementation is worse than the ANN approach at finding the next number in a sequence.

Important to note is that the neural network only finds the next number and not a formula or program like our algorithm and the methods discussed above. Furthermore, a neural network needs to be trained before it can be used to solve sequences.

7 Further Work

Further research can focus on better usage and selection of feature vectors. Our experiments used only a small number of feature vectors and because the feature vectors are the results of mathematical functions, there is a lot of room for improvement here.

8 Conclusion

Our method is able to find formulas for integer sequences with some success. The chance of success is largely dependent on the type of sequence and the relationship between the length of the given sequence and the minimal window size. The used feature vector functions also play an important role in finding formulas. The method is extensible via feature vectors which means the chance of success can be increased by using the right feature vector for the given sequence.

Our method can compete with existing methods with regard to performance, success rate, and output quality.

Acknowledgements. We would like to thank our supervisor Prof. Luc De Raedt for his guidance and support during our bachelor's thesis.

References

1. Düsel, M., Werner, A., Zeißner, T.: Solving number series with the MagicHaskeller. Technical report, University of Bamberg (2012)
2. Hernández-Orallo, J., Martínez-Plumed, F., Schmid, U., Siebers, M., Dowe, D.L.: Computer models solving intelligence test problems: progress andimplications. Artif. Intell. **230**, 74–107 (2016). https://doi.org/10.1016/j.artint.2015.09.011. http://www.sciencedirect.com/science/article/pii/S0004370215001538
3. Milovec, M.: Applying inductive programming to solving number series problems. Master's thesis, University of Bamberg (2014)
4. OEIS Foundation Inc.: The Online Encyclopedia of Integer Sequences (2018). https://oeis.org/
5. Ragni, M., Klein, A.: Predicting numbers: an AI approach to solving number series. In: Bach, J., Edelkamp, S. (eds.) KI 2011. LNCS (LNAI), vol. 7006, pp. 255–259. Springer, Heidelberg (2011). https://doi.org/10.1007/978-3-642-24455-1_24

All or In-cloud: How the Identification
of Six Types of Anomalies Is Affected
by the Discretization Method

Ralph Foorthuis[(⊠)] [ID]

UWV / Heineken, Amsterdam, The Netherlands
ralphfoorthuis@gmail.com

Abstract. Anomaly detection is the process of identifying cases, or groups of cases, that are in some way unusual and do not fit the general patterns present in the dataset. Numerous algorithms use discretization of numerical data in their detection processes. This study investigates the effect of the employed discretization method on the unsupervised detection of each of the six anomaly types acknowledged in a recent typology of data anomalies. To this end, experiments are conducted with various datasets and SECODA, a general-purpose algorithm for unsupervised non-parametric anomaly detection in datasets with numerical and categorical attributes. This algorithm employs discretization of continuous attributes, exponentially increasing weights and discretization cut points, and a pruning heuristic to detect anomalies with an optimal number of iterations. The empirical results of experiments with synthetic and real-world data demonstrate that standard SECODA can detect all six types, but that different discretization methods favor the discovery of certain anomaly types. These main findings also hold for other detection techniques using discretization.

Keywords: Anomaly detection · Outlier detection · Deviants · SECODA ·
Data mining · Typology · Discretization · Binning · Concatenation trick ·
Anomaly types · Classification

1 Introduction

The task of *anomaly detection* (AD) refers to identifying cases, or groups of cases, that are in some way unusual and do not fit the general patterns present in the dataset [1–3]. The detection of *anomalies*, which are often also referred to as outliers, deviants or novelties, is a major research topic in the overlapping disciplines of artificial intelligence [4–6], data mining [7–9] and statistics [10–12]. It is not merely of interest for academia, however, as it is also of significant value in industrial practice nowadays [13, 14, 36]. Anomaly detection can be used for discovering fraud, data quality issues, security threats, process and system failures, and deviating data points that hamper model training.

Many techniques for detecting anomalies have been devised throughout the years. The field of statistics traditionally focused mainly on parametric methods for discovering univariate outliers in each attribute (variable) separately [cf. 1, 12, 15]. Distance- and

© Springer Nature Switzerland AG 2019
M. Atzmueller and W. Duivesteijn (Eds.): BNAIC 2018, CCIS 1021, pp. 25–42, 2019.
https://doi.org/10.1007/978-3-030-31978-6_3

density-based techniques were consequently developed, allowing for non-parametric multidimensional data mining [16–18]. Another group of methods comprises complex non-parametric models, such as one-class support vector machines, ensembles and various subspace methods [19–21]. Other approaches employ reconstruction techniques or information-theoretic concepts such as entropy and Kolmogorov complexity [22, 23]. Some solutions focus on individual cases (data points) [e.g. 16, 17, 25], whereas others aim to detect groups or substructures [e.g. 8, 23]. Discretization of continuous (numerical) attributes is a technique used in many of the AD approaches, e.g. for improving accuracy and time performance of the algorithms [24–28].

SECODA is an algorithm for unsupervised non-parametric anomaly detection in datasets with continuous and categorical attributes [25, 29]. It bears similarities with, i. a., density-based AD solutions and ensembles. SECODA employs discretization of numerical attributes, exponentially increasing weights and discretization cut points, as well as a pruning heuristic to detect anomalies with an optimal number of iterations. Its rich form of discretization makes it well-suited for this paper's experimentation.

This study investigates the effect of the discretization method on the unsupervised detection of each of the six anomaly types acknowledged in a recent typology of data anomalies [3]. The experimental results not only demonstrate that SECODA, using its standard settings, is able to detect all six anomaly types, but also that different discretization methods clearly favor the discovery of different anomaly types. Moreover, the main results, as summarized in Tables 2 and 3, also hold for other techniques using discretization.

This paper proceeds as follows. Section 2 presents the necessary theoretical background. Section 3 discusses the experiments that have been conducted with several synthetic and real-world datasets. Section 4 is for conclusions.

2   Theoretical Foundations

This section presents a summary of the typology of anomalies, a brief overview of discretization theory, and an explanation of the SECODA algorithm.

2.1   Typology of Anomalies

The typology of data anomalies presented in [3] offers a theoretical and tangible *understanding* of the nature of different types of anomalies, assists researchers with systematically *evaluating* the functional capabilities of anomaly detection algorithms, and as a framework aids in *analyzing* the nature of data, patterns and anomalies. The typology uses two fundamental and data-oriented dimensions:

- *Types of Data*: The data types of the attributes that are involved in the anomalous character of a deviant case. These can be *continuous* (numerical, e.g. height or temperature), *categorical* (code- or class-based, e.g. color or blood type) or *mixed* (when both types are involved).
- *Cardinality of Relationship*: The way in which the various attributes relate to each other when describing anomalous behavior. When no relationship between the

variables exists to which the anomalous character of the deviant case can be attributed, the relationship is said to be *univariate*. It follows that the analysis can assume independence between the attributes. On the other hand, when the deviant behavior of the anomaly lies in the relationships between its variables, i.e. in the combination of its attribute values, then the relationship is said to be *multivariate*. This means the variables need to be analyzed jointly, not separately, in order to account for the relationships between them.

		Types of Data		
		Continuous attributes	**Categorical attributes**	**Mixed attributes**
Cardinality of Relationship	**Univariate**	Type I Extreme value anomaly	Type II Rare class anomaly	Type III Simple mixed data anomaly
	Multivariate	Type IV Multidimensional numerical anomaly	Type V Multidimensional rare class anomaly	Type VI Multidimensional mixed data anomaly

Fig. 1. The typology of anomalies.

These two dimensions naturally and objectively yield six basic types of anomalies. Although the typology can be used to describe aggregate anomalies (a group of cases that deviates), the focus in this study is on individual data points. The anomaly types are described below (note: the reader might want to zoom in on a digital screen to see colors, patterns and data points in detail).

- *Type I - Extreme value anomaly*: A case with an extremely high, low or otherwise rare (e.g. isolated intermediate) value for one or several individual numerical attributes. This type of outlier is typically considered in traditional univariate statistics, e.g. by using a measure of central tendency plus or minus 3 times the standard deviation or the median absolute deviation. Examples of Type I anomalies are the *Ia* and *Ib* cases in Fig. 2A.
- *Type II - Rare class anomaly*: A case with an uncommon class value for one or several individual categorical variables. Such values can be few and far between or truly unique (i.e. occur only once). An example of a Type II anomaly is the *IIa* case in Fig. 2B, which is the only square shape in the set.

- *Type III - Simple mixed data anomaly*: A case that is both a Type I and Type II anomaly, i.e. with at least one extreme value and one rare class. This anomaly type deviates with regard to multiple data types. This requires deviant values for at least two attributes, each anomalous in their own right. These can thus be analyzed separately. Analyzing the attributes jointly is not necessary because, like Type I and II anomalies, the case is not deviant in terms of a combination of values. An example of a Type III anomaly is the *IIIa* case in Fig. 2B, a unique shape at an extreme numerical position.
- *Type IV - Multidimensional numerical anomaly*: A case that does not conform to the general patterns when the relationship between multiple continuous attributes is taken into account, but that does not have extreme or isolated values for any of the individual attributes that partake in this relationship. The anomalous nature of a case of this type lies in the deviant or rare combination of its continuous attribute values. Detection therefore requires several numerical attributes that are analyzed jointly. An example of a Type IV anomaly is the *IVa* case in Fig. 2A.
- *Type V - Multidimensional rare class anomaly*: A case with a rare combination of class values. A minimum of two categorical attributes needs to be analyzed jointly to discover a multidimensional rare class anomaly. An example is this curious combination of values from three attributes used to describe dogs: 'MALE', 'PUPPY' and 'PREGNANT'. Another example is the *Va* case in Fig. 2B, which is the only red circle in the set.

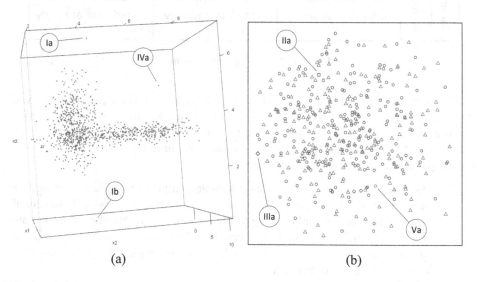

(a)   (b)

Fig. 2. (A) Mountain dataset with 3 numerical attributes; (B) ClassCircle dataset with two numerical attributes and two categorical attributes (color and shape). (Color figure online)

- *Type VI - Multidimensional mixed data anomaly*: A case with a deviant relationship between its continuous and categorical attributes. The anomalous case generally has a categorical value or a combination of categorical values that in itself is not rare in the dataset as a whole, but is only rare in its neighborhood (numerical area) or local pattern. As with Type IV and V anomalies, multiple attributes need to be jointly taken into account to identify them. In fact, multiple datatypes need to be used, as a Type VI anomaly per definition requires both numerical and categorical data. Examples of Type VI anomalies are the *VIa* cases in Fig. 6A, seemingly misplaced green cases amongst an overwhelmingly red data cloud.

The value of this typology lies not only in providing both a theoretical and tangible understanding of the types of anomalies one can encounter in datasets, but also in its ability to help evaluating which type of anomalies can be detected by a given algorithm – or a given configuration of an algorithm. See [3, 25] for more examples of anomalies.

2.2   Discretization

The task of *discretization* refers to partitioning a continuous attribute into a limited number of sub-ranges (intervals) in order to obtain a categorical data type [27, 28, 30]. Discretization is used regularly in artificial intelligence, as numerous machine learning and data mining algorithms require a categorical feature space [7, 27, 28, 30]. Examples of algorithms where discretization plays a crucial role are decision trees, random forests, Bayesian networks, naive Bayes and rule-learners. Discretization also plays an important role in anomaly detection [cf. 24, 25, 26]. Apart from the fact that techniques may require categorical data, discretization has been shown to improve the accuracy, time performance and understandability of analysis methods [27, 28, 30].

The term *arity* refers to the resulting number of intervals or partitions. Several methods allow to set this number b before running the discretization process. The range of a continuous variable is divided into intervals by $b - 1$ cut points. An individual *cut point* or split point is a real value at the position where an interval boundary is located, dividing the range into two intervals.

Discretization methods can be supervised, taking into account the training set's class label that ultimately needs to be predicted, or unsupervised, thus not taking into account a dependent variable. Two main unsupervised discretization methods exist, both of them often referred to as *binning* [7, 26, 27, 31]. *Equiwidth* discretization refers to equal interval binning. This method divides the range of an attribute's observed continuous values into b bins of the same value interval. The second method is *equidepth* discretization, which refers to equal frequency binning and divides a continuous attribute into b bins that each contain the same number of cases. In both methods b is provided as input to the discretization function. The two discretization techniques have been used for anomaly detection [e.g. 24, 25, 26].

Discretization methods can be characterized in several ways [28, 30, 31]. Binning techniques can be global or local, albeit both unsupervised methods employed in this study are global. This means that they use the entire value space for partitioning, independently of the characteristics of local regions. Methods can also be direct or incremental, with the latter referring to techniques that pass through the data several times to arrive at an optimal discretized attribute. The equiwidth and equidepth methods are direct, meaning that they require only one pass. Finally, both binning methods discretize the data for each attribute separately, so these binning solutions do not take into account any relationships between the variables.

2.3 SECODA

SECODA, an abbreviation for segmentation- and combination-based detection of anomalies, is a general-purpose algorithm for unsupervised anomaly detection in datasets with mixed data [25, 29, 40]. The algorithm is non-parametric in nature and therefore does not assume any specific data distribution. It investigates the joint distribution to discover high-density patterns and low-frequency deviations in the dataset, taking into account any relationship that may exist between the attributes. To this end, SECODA iteratively searches the dataset until the cases have been scrutinized with sufficient detail.

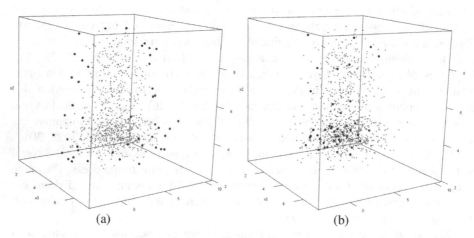

Fig. 3. (A) The large black dots represent the top 45 anomalies of the Mountain set resulting from equiwidth binning; (B) The top 45 anomalies from equidepth binning.

SECODA is guaranteed to identify cases with unique or very rare combinations of attribute values. The algorithm uses the histogram-based approach to assess the density of each combination (or "constellation") of categorical and continuous attribute values. The concatenation trick, which combines categorical and discretized continuous attributes into a new constellation feature, is used to analyze different data types in a joint fashion. In conjunction with recursive binning this captures complex relationships between attributes. In subsequent iterations SECODA uses increasingly narrow discretization intervals in order to add more detail and precision to the analysis and identify more subtle anomalies. The distance between data points in numerical space is implicitly accounted for by this iterative binning process. A pruning heuristic as well as exponentially increasing weights and arity are employed to speed up the analysis. The increasing arity (providing more localized details) and weights (allowing for optimally combining the results obtained from different iterations) also help to avoid discretization error and detection bias.

Note that recursive discretization is not employed by SECODA to find a single, optimal value for the arity parameter b, because it exploits the information from all binning iterations. Put differently, SECODA is an algorithm that recursively collects and uses the information from a discretization method that is itself applied in each iteration in a direct (instead of incremental) manner. The input parameter b is thus not provided by the user, but repeatedly by SECODA until a stopping criterion is reached.

The SECODA approach has several favorable properties. It is a relatively simple algorithm that does not require expensive point-to-point calculations. Only basic data operations are used, making it suitable for sets with large numbers of rows and for in-database analytics and machines with relatively little memory. The algorithm scales linearly with dataset size, and for extremely large sets a longer computation time is hardly required because additional iterations would not yield a meaningful gain in precision. An exploratory AD analysis could limit the runtime by simply setting a maximum of e.g. 10 iterations, which for very large sets can be expected to be faster than point-to-point algorithms. The technique can also easily be implemented for parallel processing architectures. All kinds of relationships between attributes are taken into account, such as (non)linear associations, interactions, collinearity and relations between variables of different data types. Although SECODA is vulnerable to the curse of dimensionality, general techniques such as feature bagging and random projection can be applied to deal with this. Missing values are automatically handled as one would functionally desire in an AD context, with only very rare missing values being considered anomalous. Finally, the pruning heuristic is a self-regulating mechanism during runtime, dynamically deciding how many cases to discard. After converging the algorithm returns a score vector so that each case gets assigned a degree of anomalousness, with lower scores representing more deviant occurrences.

SECODA has been evaluated in an academic context and has been used in practice as well to discover anomalies in the Polis Administration, an official register maintaining masterdata regarding the salaries, social security benefits, pensions and income relationships of people working or living in the Netherlands [25, 37, 40]. The evaluation involved applying the algorithm to various synthetic and real-world datasets. Using ROC and PRC curves, as well as AUC and partial AUC metrics, it was demonstrated that this AD solution is able to successfully detect a wide variety of anomaly types. It has also been shown that the algorithm has low memory requirements and scales linearly with dataset size. SECODA has not been tested on all six types of anomalies, as the full typology was published later. Section 3 will demonstrate that the algorithm is indeed able to detect all types, and is therefore well-suited for experiments studying the effects of discretization on the detection of these types.

SECODA can be downloaded for free as a package for the R environment (see Remarks). The implementation offers various options, such as the minimum and maximum number of iterations, a pruning parameter, and the iteration after which the heuristics should start to run. These options generally have trivial consequences and are mainly intended to tweak the amount of analysis detail and running time, so the standard settings normally suffice. This is desirable because algorithms for data mining are ideally parameter-free in order to discover the true patterns and deviations in a simple and objective fashion [23, cf. 18]. On the other hand, however, it is widely acknowledged that the world – and therefore the datasets that it produces – is extremely complex, and that no single algorithm or algorithm setting is thus able to perform excellent in all situations [18, 32–34]. This also holds in the context of anomaly detection [35, 36] and discretization [30]. Section 3 therefore investigates the effect of the binning method, another parameter that the analyst can set before running SECODA, on detecting the different types of anomalies defined in Sect. 2.1.

3   Empirical Experiments

3.1   Research Design and Datasets

This study uses several synthetic and real-world datasets to investigate whether and how the discretization method affects the detection of the various anomaly types. The simulated datasets are labelled, which makes them suitable for verifying whether AD algorithms can readily detect the anomalies. The real-world dataset, drawn randomly from the aforementioned Polis Administration and anonymized subsequently, is unlabeled. The sets are described in Table 1 and are visually depicted in Figs. 2, 3, 4, 5, 6 and 7. See the Remarks for download options. The R environment 3.4.3, RStudio 1.1.383, SECODA 0.5.3 and rgl 0.98.22 were used to generate the synthetic datasets and conduct the experiments. SECODA's heuristics for speeding up the analysis (e.g. pruning, which in a standard configuration starts being applied after 10 iterations) were not used in order to ensure maximum precision of the results.

Table 1. Datasets used for experiments.

Dataset	Nature	Datatypes	# Cases	Types of anomaly
ClassCircle	Simulated	2 num, 2 categ	422	Type II, III, V
Mountain	Simulated	3 numerical	943	Type I, IV
NoisyMix	Simulated	3 num, 2 categ	3867	Type II, VI
Sword	Simulated	2 num, 1 categ	7024	Type II, III, VI
Helix	Simulated	3 num, 1 categ	1410	Type I, IV, VI
Polis dataset	Real-world	3 num, 1 categ	304726	Type I, II, IV, VI

Although the multivariate anomaly types can be used to describe aggregate anomalies – i.e. a group of related cases that deviates as a whole [3] – this study will focus solely on deviants that are atomic, single cases in independent data. The reason for this is that detecting grouped anomalies generally requires special-purpose approaches.

3.2 Results and Discussion

In the first series of experiments the five simulated datasets were used to study whether SECODA was able to identify the six types of anomalies presented in Sect. 2.1. Note that [25] was not able to evaluate the algorithm on all six types because the full typology of [3] had not been developed at the time. The standard configuration of SECODA employs equiwidth binning and was indeed able to detect all types of anomalies. The subsequent series of experiments involved running SECODA with the non-standard equidepth setting to investigate what types of anomalies were identified in this fashion and how this compared to equiwidth AD.

With regard to a univariate analysis of a *single numerical attribute*, it is evident that the equiwidth setting is the preferred and basically only sensible option. This setting is able to detect isolated Type I cases, both extremely large or small values and rare intermediate data points. The equidepth setting, even though many discretization iterations were generally required before converging, was not able to detect these obvious anomalies and resulted in all cases getting a very high and non-discriminating score. This can be easily explained by the nature of equidepth discretization, since every bin gets assigned the same number of cases (although slight differences in frequency might occur if the set cannot be split evenly). SECODA's repeated binning with increasingly narrow intervals does not change this fact.

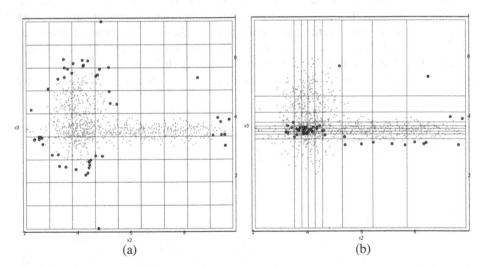

(a)   (b)

Fig. 4. Two attributes of the Mountain set analyzed with (A) equiwidth binning and (B) equidepth binning. It is clear the bins (intervals) are very different.

For the Mountain set with *multiple numerical attributes* the equiwidth setting was also found to be the superior choice, as it was readily able to detect the 3 labelled Type I and IV anomalies. Furthermore, the other cases with a low score also made sense, as they were all relatively isolated cases at the fringe of the data cloud. With the equidepth setting only 1 of the 3 labelled anomalies were detected (the Type IV case of

Fig. 2A). In addition, most of the other low-score equidepth results were positioned in the middle of the data cloud, seemingly without a good reason why these should be considered more anomalous than other data points. The difference between the two binning methods is illustrated by Fig. 3A on the left depicting the 45 most anomalous cases found by equiwidth binning, which are mostly outlying and include the 3 labelled anomalies, and Fig. 3B showing the 45 lowest-score cases, which are mainly positioned in the high-density center of the data cloud. (Note that the aforementioned 3 true anomalies, which can be clearly seen in Fig. 2A, are not visible from this angle.)

Figure 4 illustrates the difference between the two discretization methods even more clearly, showing also how the cut points result in very different constellations (multi-dimensional segments of the data). Note that the cut points and resulting constellations are the result of a single discretization run, and thus only illustrate a part of the AD process. The large black dots represent the top 50 anomalies identified by the algorithm after 10 (equiwidth) and 13 (equidepth) iterations, analyzing two of the set's attributes. The equiwidth AD results represent the most isolated points, often at the fringes of the data cloud, and as such make sense. The equidepth run detects the Type IV anomaly as one of the most extreme cases, but also yields many meaningless false positives at the center of the cloud. This will be explained in more detail later.

When disregarding the categorical attributes in the Helix and NoisyMix sets, the results are very similar. Type IV anomalies can be detected relatively well by equidepth binning, albeit with more false positives. Type I deviants are not detected, although they may be found if they have extreme values for multiple numerical attributes and thus are anomalous with regard to the combination of these values. Figure 5 illustrates a single equidepth binning iteration for the NoisyMix set, with large black dots representing the identified anomalies. Due to the slightly different data distribution the lowest-score cases are positioned around the dense data cloud, rather than at the center of it.

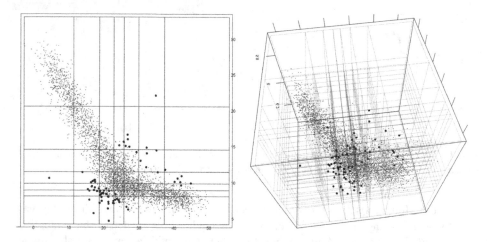

Fig. 5. Equidepth binning for the NoisyMix shown in 2D and 3D.

In short, the equidepth setting is most certainly not suitable for AD analysis of univariate numerical vectors (hosting Type I cases) and is reasonably equipped for dealing with multivariate numerical sets (hosting Type IV cases). Equiwidth binning yields more meaningful results as it directly targets the numerically isolated cases.

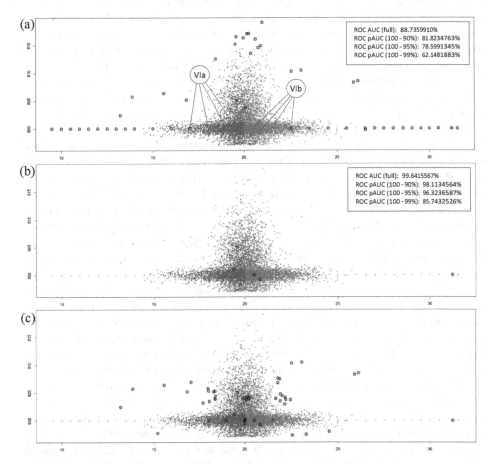

Fig. 6. From top to bottom: (A) The top 50 anomalies (black circles) of the Sword set from equiwidth binning; The (B) top 5 and (C) top 50 anomalies from equidepth binning. (Color figure online)

When analyzing a dataset containing *only categorical attributes*, the discretization method does not in any way influence the results. This is entirely to be expected, as discretization of continuous data should not affect a purely categorical analysis. The binning method provided by the analyst as an input parameter to the algorithm is simply irrelevant in this situation. Tests on several datasets indeed confirm this when running the algorithm with the two settings. In sets with *mixed data* both numerical and categorical attributes are present, and the returned scores of the two discretization methods can be expected to be different. However, the effect depends on the type of anomaly and the distribution of the data. Truly unique Type II or III univariate class-based anomalies will be recognized as unique, regardless of the binning method, and get assigned the lowest score possible. The same holds for unique combinations of classes, i.e. Type V cases. Experiments with the datasets that contain categorical data confirmed this as well, with SECODA returning the lowest anomaly score for such unique cases with both methods. However, when the Type II, III or V anomalies are rare in the dataset (rather than truly unique), the numerical data may influence the score. This can be expected because the rare cases can be close or distant neighbors and also 'compete' with e.g. very isolated Type I and VI deviants. However, regardless of the binning method one would still expect these anomalies to be identified, returning relatively low anomaly scores for such cases. This is confirmed as well, although with some interesting differences between the two discretization methods (see below).

The binning method possibly has the most interesting impact on the detection of Type VI anomalies. These do not feature truly (globally) unique classes, because these classes are common in other areas of the numerical space. The detection of these local anomalies may therefore very well be affected by the discretization technique, an expectation that was confirmed by the experiments. In several datasets it was observed that equidepth binning often yields superior results when the goal is to detect Type VI anomalies. This is illustrated by Fig. 6A at the top, where it can clearly be seen that the equiwidth analysis results in a variety of anomalies. However, due to the nature of the Sword dataset, which contains many numerically isolated cases, most of the top 50 anomalies are Type I and IV outliers. The Type II and III anomalies are also detected, but the Type VI anomalies less so. The equidepth analysis presented in Fig. 6B and C results in quite different cases being denoted as most extreme anomalies. It can be seen that the top 5 cases are mostly Type VI anomalies, which are located in dense (rather than sparsely populated) regions of the space. The Type II case at height 805 and the Type III case at the far right of Fig. 6B are truly unique classes due to their color and are therefore regarded as highly anomalous by both binning methods. Rare (as opposed to truly unique) Type II and V anomalies, which can but do not have to be isolated, are also detected more readily with equidepth binning when not located in low-density areas. Equiwidth binning will acknowledge a handful of neighboring rare classes (i.e. a very small 'cluster') as moderately anomalous, regardless of whether they lie inside or outside the data cloud. This is due to the fact that they are not truly unique. Equidepth binning, on the other hand, will recognize them as highly anomalous if they lie within the cloud, but not if they lie outside it (see the five detected purple cases in the middle of Fig. 6C). Figure 6 also shows the ROC AUC and 3 specificity partial AUCs for the specific task of detecting the in-cloud high-density anomalies (not the numerically isolated cases). In short, equiwidth discretization is well-equipped for detecting all anomaly types, including isolated occurrences. Equidepth binning, although more vulnerable to yielding false positives, is relatively well-equipped for detecting Type VI and in-cloud Type II and V anomalies.

To further investigate these findings, SECODA was used to analyze a sample from the aforementioned real-world Polis dataset. A similar effect was observed here. Figure 7A on the left illustrates the results of AD with equiwidth binning, which yielded a wide variety of anomalies, including many isolated cases. Figure 7B shows the results of AD with equidepth discretization, with the most extreme anomalies found to be positioned in the center's high-density area. Both figures also have a zoomed-in view at the bottom, where the difference can be seen in more detail for each binning method.

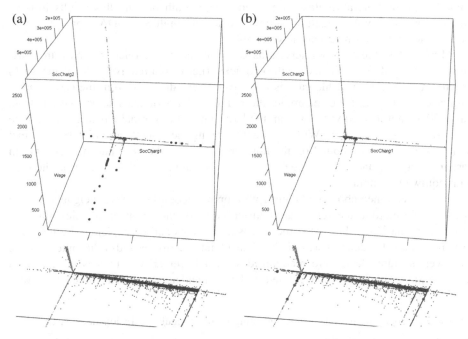

Fig. 7. (A) The large dots represent the 40 most extreme anomalies of the Polis set detected by equiwidth binning (the bottom is zoomed-in); (B) The top 40 anomalies from equidepth binning

At this point it is valuable to discuss the reasons why equiwidth and equidepth discretization yield different results in an AD context. In general, equiwidth binning performs better in terms of overall functional performance, i.e. the capability to detect a wide variety of meaningful anomalies. The reason for this is that equiwidth binning (or at least a single binning run with only one value for b) uses fixed value intervals, resulting in isolated Type I, III and IV cases to be placed in near empty bins. This also holds when an algorithm such as SECODA repeatedly discretizes the continuous attributes using many values for b during the analysis, thus creating few bins in early iterations and many bins in later iterations (the recursive binning ensures that more distant anomalies get lower scores). It is known from the literature that data analysis with equiwidth binning is sensitive to outliers, a property that is usually seen as a disadvantage [7, 27, 28, 31]. However, in the context of anomaly detection this sensitivity can be exploited, resulting in relatively easy detection of sparse data by isolating them in separate bins. Equidepth binning, on the other hand, fails this detection of isolated cases, since the value ranges of the bins are stretched so as to fill them with an equal amount of data points. For example, in a typical Gaussian distribution the discretization intervals at the tails will be very wide because these regions are sparsely populated and the bins have to be filled with a given amount of cases. Moving inwards to the mean of the Gaussian distribution the bins will get narrower. The consequence is that univariately numerically isolated cases are not detected, as the focus is then on the categorical abnormality and – in case of multivariate analysis – on the combination of values from multiple attributes. Compared to equiwidth binning this results in (univariate) low-density cases getting assigned relatively high SECODA scores and non-isolated deviant cases relatively low scores.

Moreover, in the narrow intervals used for univariate high-density areas, the class values of Type II, V or VI cases will be quickly (i.e. with a relatively low value for b) unique in its bin, even if the case is not located very distantly from the cases with a similar color. In Fig. 6, for example, the red Type VI anomalies somewhat left from value 20 are not located very far from the large amount of normal red cases that can be seen from value 20 and up. With equidepth binning the discretization intervals in that region of the variable plotted on the horizontal axis will be very narrow, resulting in earlier separation and therefore detection of these anomalies than would be the case with equiwidth binning.

It was mentioned above that equidepth binning recognizes several neighboring rare classes (referred to as the very small 'cluster') as very anomalous if they lie within the rest of the data cloud, but not if they lie isolated. This can be explained by the same reason: within the dense parts of the data cloud the discretized value intervals are narrower, so rare classes are recognized with a lower arity b than with equiwidth discretization. This means they get detected earlier and are denoted as more anomalous.

Table 2. Underlying reason for discriminating between regular and anomalous data points.

Underlying reason for discriminating	EW	ED
Categorical data with an unbalanced class distribution is present	✔	✔
Bins get scarcely filled with isolated points	✔	×
Combinations of numerical and/or categorical values yield infrequent occurrences	✔	✔

Equidepth binning in an AD context thus scrutinizes the dense regions of the distribution in more detail (if these regions can be detected univariately). This method of discretization disregards tail and intermediate data points that are isolated in numerical space. Instead, when a multivariate analysis is conducted, the focus will be on *uncommon class values* and *rare combinations* of (continuous and categorical) attribute values. Equidepth discretization thus ignores univariately isolated cases and, more so than equiwidth analysis, has a propensity to detect anomalies that lie amongst other data points. It favors detecting cases that w.r.t. numerical attributes are located in univariate high-density regions. The discretization process, which handles individual attributes, will place cases that are located in univariate high-density regions in very thin univariate bins, i.e. in narrow value intervals. If these cases are also located in the univariate high-density ranges of other attributes, the multidimensional intersection will thus yield relatively low-density, sparsely populated constellations. In a purely numerical dataset this property will denote as anomalies both Type IV cases (true deviants) and points in or around the densest areas of the data cloud (often false positives, but sometimes interesting subtle deviants). Figure 4B clearly illustrates this mechanism, by showing that very dense regions have very small segments or constellations, which then may happen to contain relatively few cases. These arguments holds both for one-time discretization and the iterative binning of SECODA. For mixed data this works well for discovering Type VI anomalies, as well as for Type II and V cases located in high-density areas.

Table 2 succinctly states why equidepth binning can discriminate between normal and anomalous cases. It is not because *bins get scarcely filled with isolated points*, because all bins are filled with an equal amount of cases. Rather, it is because *categorical data with an unbalanced class distribution* is present or because the *combination of numerical and/or categorical values* yields infrequent occurrences. Equiwidth binning, on the other hand, utilizes all these three discriminating properties.

Table 3 summarizes the findings for each anomaly type. The *Impact?* column refers to whether there exists a direct impact of using equidepth (ED) instead of equiwidth (EW) binning for the anomaly type. The *Useful?* column denotes whether equidepth binning can be useful in some situations for detecting the given type.

Table 3. Impact of discretization method on detection of anomaly types.

Type	Impact?	Useful?	Explanation
I	Y	N	ED cannot discriminate between the univariate numerical values and is intrinsically not equipped to detect this type
II	N/Y	Y	ED is identical to EW when analyzing a single categorical attribute. It can be more useful than EW if the goal is to detect (non-unique) rare Type II anomalies in numerically high-density regions in an analysis of mixed data
III	Y	Y	ED detects truly unique classes equally well as EW, but the latter shows slightly better performance with rare classes (because EW will exploit their isolated position better)
IV	Y	Y	ED detects many Type IV cases, but also yields more false positives and false negatives, and is thus not optimally equipped to detect this type. ED can sometimes detect more subtle Type IV cases at dense areas than EW can
V	N/Y	Y	ED is identical to EW in a set with merely categorical data. It can be more useful than EW if the goal is to detect (non-unique) rare Type V anomalies in numerically high-density regions in an analysis of mixed data
VI	Y	Y	ED tends to favor the detection of Type VI anomalies and can be more useful than EW if this is indeed the goal

These main conclusions also hold for single, non-iterative binning operations, e.g. using only 7 intervals to discretize each continuous attribute. However, the recursive binning of SECODA accounts for the distance between data points and is thus able to take this into account to calculate the degree of deviation. A single discretization run, on the other hand, requires the analyst to pick an arbitrary number of bins and cannot return such information on the degree of anomalousness as a result of this rather crude form of binning.

As a final note, equidepth discretization can be useful in practical situations, as it is known that in some settings it is valuable to detect non-isolated and relatively subtle deviations rather than cases that are extreme and rare on all accounts [cf. 38, 39].

4   Conclusion

This study has analyzed the impact of two discretization methods on the detection of different types of anomalies. The SECODA algorithm was used in the experiments because of its rich form of discretization. The empirical results of the analysis with synthetic and real-world data demonstrate that discretization, including its employment in the standard SECODA algorithm, can be used to detect all six types of anomalies. However, the equiwidth and equidepth discretization techniques yield notably different results and favor the discovery of certain anomaly types. Equiwidth and equidepth SECODA can therefore best be seen as two different algorithms. Equiwidth SECODA is a general-purpose algorithm, whereas the equidepth version is a special-purpose technique focusing on specific anomaly types. The main conclusions, as summarized in Tables 2 and 3, also hold for techniques that perform discretization only once, although the results hereof will be less precise and will not account for the distance between data points.

In general, if the analyst does not know beforehand in what type of anomaly he or she is interested, then equiwidth discretization is the preferred option. This will conduct a general-purpose anomaly detection and ensure that all anomaly types will be detected. If on the other hand the focus is on identifying anomalies that are not located in extreme or isolated regions of the numerical space, equidepth discretization should be used. The equidepth binning option favors the detection of Type VI anomalies as well as Type II and V cases that are found inside data clouds rather than in sparsely populated regions.

Remarks A SECODA implementation, various datasets and the code to analyze them in R can be downloaded from www.foorthuis.nl (see the *SECODA resources for R* section).

Final version: September 8th 2019. This research was sponsored in part by UWV (Uitvoeringsinstituut Werknemersverzekeringen).

This publication is an extended version of the BNAIC 2018 paper *The Impact of Discretization Method on the Detection of Six Types of Anomalies in Datasets.*

References

1. Barnett, V., Lewis, T.: Outliers in Statistical Data, 3rd edn. Wiley, Chichester (1994)
2. Goldstein, M., Uchida, S.: A comparative evaluation of unsupervised anomaly detection algorithms. PLoS ONE **11**(4), e0152173 (2016)

3. Foorthuis, R.: A typology of data anomalies. In: Medina, J., et al. (eds.) IPMU 2018. CCIS, vol. 854, pp. 26–38. Springer, Cham (2018). https://doi.org/10.1007/978-3-319-91476-3_3
4. Pang, G., Cao, L., Chin, L.: Outlier detection in complex categorical data by modelling the feature value couplings. In: Proceedings of the 25th International Joint Conference on Artificial Intelligence (2016)
5. Riahi, F., Schulte, O.: Propositionalization for unsupervised outlier detection in multi-relational data. In: Proceedings of the 29th International Florida Artificial Intelligence Research Society Conference (2016)
6. Hengst, F., den Hoogendoorn, M.: Detecting interesting outliers: active learning for anomaly detection. In: Proceedings of the 28th Benelux Conference on Artificial Intelligence, Amsterdam, The Netherlands (2016)
7. Tan, P., Steinbach, M., Kumar, V.: Introduction to Data Mining. Addison-Wesley, Boston (2005)
8. Noble, C.C., Cook, D.J.: Graph-based anomaly detection. In: Proceedings of the Ninth ACM SIGKDD International Conference on Knowledge Discovery and Data Mining (2003)
9. Schubert, E., Weiler, M., Zimek, A.: Outlier detection and trend detection: two sides of the same coin. In: Proceedings of the 15th IEEE International Conference on Data Mining Workshops (2015)
10. Hubert, M., Rousseeuw, P., Segaert, P.: Multivariate functional outlier detection. Stat. Methods Appl. **24**(2), 177–202 (2015)
11. Ranshous, S., Shen, S., Koutra, D., Harenberg, S., Faloutsos, C., Samatova, N.F.: Anomaly detection in dynamic networks: a survey. WIREs Comput. Stat. **7**(3), 223–247 (2015)
12. Fielding, J., Gilbert, N.: Understanding Social Statistics. Sage Publications, London (2000)
13. Gartner: Hype Cycle for Data Science and Machine Learning, 2017. Gartner, Inc (2017)
14. Forrester: The Forrester Wave: Security Analytics Platforms, Q1 2017. Forrester Research, Inc. (2017)
15. Leys, C., Ley, C., Klein, O., Bernard, P., Licata, L.: Detecting outliers: do not use standard deviation around the mean, use absolute deviation around the median. J. Exp. Soc. Psychol. **49**(4), 764–766 (2013)
16. Knorr, E.M., Ng, R.T.: Algorithms for mining distance-based outliers in large datasets. In: VLDB 1998, Proceedings of the 24th International Conference on Very Large Data Bases (1998)
17. Breunig, M.M., Kriegel, H., Ng, R.T., Sander, J.: LOF: identifying density-based local outliers. In: Proceedings of the ACM SIGMOD Conference on Management of Data (2000)
18. Campos, G.O., et al.: On the evaluation of unsupervised outlier detection: measures, datasets, and an empirical study. Data Min. Knowl. Discovery **30**(4), 891–927 (2016)
19. Schölkopf, B., Williamson, R., Smola, A., Shawe-Taylor, J., Platt, J.: Support vector method for novelty detection. Adv. Neural Inf. Process. **12**, 582–588 (2000)
20. Liu, F.T., Ting, K.M., Zhou, Z.: Isolation-based anomaly detection. ACM Trans. Knowl. Discov. Data **6**(1), 3 (2012)
21. Shyu, M.L., Chen, S.C., Sarinnapakorn, K., Chang, L.W.: A novel anomaly detection scheme based on principal component classifier. In: Proceedings of the ICDM Foundation and New Direction of Data Mining workshop, pp. 172–179 (2003)
22. Pimentel, M.A.F., Clifton, D.A., Clifton, L., Tarassenko, L.: A review of novelty detection. Signal Process. **99**, 215–249 (2014)
23. Keogh, E., Lonardi, S., Ratanamahatana, C.A.: Towards parameter-free data mining. In: Proceedings of the Tenth ACM SIGKDD International Conference on Knowledge Discovery and Data Mining, Seattle (2004)

24. Goldstein, M., Dengel, A.: Histogram-based outlier score (HBOS): a fast unsupervised anomaly detection algorithm. In: Proceedings of the 35th German Conference on Artificial Intelligence (KI-2012), pp. 59–63 (2012)
25. Foorthuis, R.: SECODA: segmentation- and combination-based detection of anomalies. In: Proceedings of the 4th IEEE International Conference on Data Science and Advanced Analytics (DSAA 2017), pp. 755–764, Tokyo (2017). https://doi.org/10.1109/dsaa.2017.35
26. Aggarwal, C.C., Yu, P.S.: An effective and efficient algorithm for high-dimensional outlier detection. VLDB J. **14**(2), 211–221 (2005)
27. Dougherty, J., Kohavi, R., Sahami, M.: Supervised and unsupervised discretization of continuous features. In: Proceedings of the Twelfth International Conference on Machine Learning (1995)
28. Kotsiantis, S., Kanellopoulos, D.: Discretization techniques: a recent survey. GESTS Int. Trans. Comput. Sci. Eng. **32**, 47–58 (2006)
29. Foorthuis, R.: Anomaly detection with SECODA. In: Poster Presentation at the 4th IEEE International Conference on Data Science and Advanced Analytics (DSAA), Tokyo (2017). https://doi.org/10.13140/rg.2.2.21212.08325
30. Yang, Y., Webb, G.I., Wu, X.: Discretization methods. In: Maimon, O., Rockach, L. (eds.) Data Mining and Knowledge Discovery Handbook. Kluwer Academic Publishers (2005)
31. Li, H., Hussain, F., Tan, C.M., Dash, M.: Discretization: an enabling technique. Data Min. Knowl. Disc. **6**(4), 393–423 (2002)
32. Wolpert, D.H., Macready, W.G.: No Free Lunch Theorems for Search. Technical report SFI-TR-95-02-010, Santa Fe Institute (1996)
33. Clarke, B., Fokoué, E., Zhang, H.H.: Principles and Theory for Data Mining and Machine Learning. Springer, New York (2009). https://doi.org/10.1007/978-0-387-98135-2
34. Rokach, L., Maimon, O.: Data Mining With Decision Trees: Theory and Applications, 2nd edn. World Scientific Publishing, Singapore (2015)
35. Janssens, J.H.M.: Outlier Selection and One-Class Classification. Ph.D. thesis, Tilburg University (2013)
36. Maxion, R.A., Tan, K.M.C.: Benchmarking anomaly-based detection systems. In: International Conference on Dependable Systems and Networks, New York (2000)
37. LAK: Anomaly Detection at the Dutch Alliance on Income Data and Taxes (2018). www.loonaangifteketen.nl
38. Pijnenburg, M., Kowalczyk, W.: Singular outliers: finding common observations with an uncommon feature. In: Medina, J., Ojeda-Aciego, M., Verdegay, J.L., Perfilieva, I., Bouchon-Meunier, B., Yager, R.R. (eds.) IPMU 2018. CCIS, vol. 855, pp. 492–503. Springer, Cham (2018). https://doi.org/10.1007/978-3-319-91479-4_41
39. Greenacre, M., Ayhan, H.: Identifying Inliers. Barcelona GSE Working Paper Series (2014)
40. Foorthuis, R.: (Un)certain anomalies in income data. In: Presentation at the Mini-Symposium on Uncertainty in Data-Driven Systems, Utrecht University, 28 January 2019

Topic Modeling for Exploring Cancer-Related Coverage in Journalistic Texts

Naomi Hariman[1,2(✉)], Marjolein de Vries[1,3], and Ionica Smeets[1]

[1] Science Communication and Society, Faculty of Science, Leiden University,
Leiden, The Netherlands
i.smeets@biology.leidenuniv.nl
[2] Bio-Pharmaceutical Sciences, Leiden University, Leiden, The Netherlands
[3] Mathematics and Computer Science, Eindhoven University of Technology,
Eindhoven, The Netherlands

Abstract. Topic modeling has been used for many applications, but has not been applied to science and health communication research yet. In this paper, using topic modeling for this novel domain is explored, by investigating the coverage of cancer in news items from the New York Times since 1970 with the Latent Dirichlet Allocation (LDA) model. Content analysis of cancer in print media has been performed before, but at a much smaller scope and with manual rather than computational analysis. We collected 45.684 articles concerning cancer via the New York Times API to build the LDA model upon.

Our results show a predominance of breast cancer in news articles as compared with other types of cancer, similar to previous studies. Additionally, our topic model shows 6 distinct topics: research on cancer, lifestyle and mortality, the healthcare system, business and insurance issues regarding cancer treatment, environmental politics and American politics on cancer-related policies.

Since topic modeling is a computational technique, the model has more difficulty with understanding the meaning of the analyzed text than (most) humans. Therefore, future research will be set up to let the public contribute to analysis of a topic model.

Keywords: Topic modeling · Cancer · Content analysis

1 Introduction

Covering cancer accurately in news media comes with unique constraints due to the disease's highly complex, technical and emotional nature [1]. Early content analysis studies starting from the 1970s provided some of the first insights into how cancer was reported in a variety of American newspapers [1] (n = 2.138) [2] (n = 1.466). These studies showed that cancer news stories did not generally address a specific type of cancer, and articles that did focus on a particular type of cancer mostly focused on breast and lung cancer.

N. Hariman and M. de Vries—Contributed equally to this paper.

© Springer Nature Switzerland AG 2019
M. Atzmueller and W. Duivesteijn (Eds.): BNAIC 2018, CCIS 1021, pp. 43–51, 2019.
https://doi.org/10.1007/978-3-030-31978-6_4

A similar content analysis study was later conducted by Clarke and Everest, who analyzed articles about cancer ($n = 131$) in a wide range of magazines published between 1991 and 2001 [3]. Their results also showed a predominance of articles addressing breast cancer as compared with other types of cancer, even though breast cancer did not have the highest incidence rate at that time. In addition, Clarke and Everest focused on the portrayal of cancer in their articles, and found that an emphasis on fear and the usage of battle metaphors were often used when framing cancer.

Moreover, Musso and Wakefield [4] analyzed cancer coverage in Canadian newspapers from 2003 to 2004 ($n = 464$), and also found that if an article addressed a specific type of cancer, 40% of the cases it was breast cancer and 21% of the time it was lung cancer, even though lung cancer had the highest incidence rate among all Canadians at the time. Analysis of latent themes showed that the newspapers addressed the relation of lifestyle and personal habits, such as smoking, and cancer the most, as well as the effect of environmental exposures and biological factors on cancer and issues within the healthcare system.

The aforementioned studies reveal how cancer is framed and covered in newspapers and magazines. However, as the articles were analyzed manually, researchers were left with some restrictions such as: a large amount of content to analyze but a short timeframe [1, 2], or a longer timeframe but a lower amount of content to analyze [3, 4]. The use of computational techniques gets rid of such restraints as content analysis can now be done in an automatic or semi-automatic manner. With the rise of the internet, more and more information is available online, and a large set of news items covering cancer can be acquired quite easily. The New York Times, for example, provides an application programming interface (API) for access to their articles from the current days back to 1851 [5]. Moreover, with automatic techniques such as topic modeling, large amounts of texts can be analyzed automatically instead of manually. Consequently, it is now possible to investigate the coverage of cancer in journalistic texts with both a long timeframe as well as a large amount of content to analyze. Therefore, in this paper, we use topic modeling to investigate the coverage of cancer in news items from the New York Times since 1970.

Topic modeling has already been used in a wide range of applications, from analyzing tweets on Twitter to find trending topics [6, 7] or a user's political orientation [8], to analyzing Facebook messages and the relation of resulting topics with personality, age and gender [9] and recommending scientific articles [10]. Even though topic modeling has been widely studied within the computational linguistics field, it has not been used to answer questions within science and health communication yet. Jacobi, van Atteveldt and Welbers [11] have demonstrated the usefulness of topic modeling for journalism research, by exploring the framing of the issue 'nuclear power' in the New York Times since 1945. Accordingly, our paper will make a first step into applying topic modeling to the science and health communication research domain.

2   Method

In this section, we discuss the data collection (Sect. 2.1), data pre-processing (Sect. 2.2) and the use of the Latent Dirichlet Allocation (LDA) model for topic modeling (Sect. 2.3).

2.1   Data Collection

The New York Times Article Search API python wrapper *NYTimesArticleAPI* [5] was used to retrieve the data. All articles containing the query 'cancer', 'chemotherapy', 'malignant', 'carcinogen' and 'tumor' were collected. The begin date of the search was set to January 1, 1970, in line with previous research [1, 2]. For each article, we retrieved the main headline and snippet (i.e. often the first paragraph of the news item), and pasted them together as one document for further analysis. In total, 57.402 articles were retrieved.

2.2   Data Pre-processing

The documents were stripped of punctuation, numbers and other special characters (e.g. :, . @ ' ' -)[1]. Words containing hyphens such as 'cancer-like' were split into two, e.g. 'cancer' and 'like'. Then, stop word removal was performed using the English stop words list from the *nltk.corpus* package [12]. Stop words in the list that contained hyphens were also split into two before performing stop word removal. As the LDA model cannot work well with small documents [13], any document that contained 20 words or less after stop word removal was removed from the dataset. These procedures resulted in 45.684 documents remaining for use in the LDA model. All documents were then shuffled to prevent from time bias in the model.

Subsequently, the documents were tokenized and bigrams and trigrams were created to group together words that often occurred together in the documents, by using the *models.Phrases* function from the *gensim* package [14]. If a word combination occurred at least 20 times in the entire dataset and had a *models.Phrases* threshold score of minimally 100, a bigram or trigram was made.

Next, lemmatization was performed on the dataset using the *spacy* model for English [15]. Lemmatization was performed on nouns, adjectives, verbs and adverbs (detected with Part-Of-Speech-tags), and a word was removed if it did not fall into one of these four categories. After lemmatization, terms were removed if they met one of the following criteria: having a document frequency of lower than 10, occurring in more than 25% of all the documents, or a character length of less than 3 characters. Additionally, terms that occurred less than 17 times in the entire corpus were removed. The terms 'immunotherapy' and 'cancer causing' were manually re-added, since they had been removed during this procedure but were deemed important and relevant to the research. As a result, the dictionary contained 7.312 unique terms.

---

[1] Python code can be found at github.com/NHariman/LDA-model-SCS-2018.

2.3   Latent Dirichlet Allocation

In our work, we use the Latent Dirichlet Allocation (LDA) [16] model for topic modeling. The LDA model is a generative probabilistic model that represents documents as random mixtures over latent topics. Each topic is then characterized by a distribution over words. Our LDA model was created using the *models.ldamodel* function from the *gensim* package [14]. The model was updated every term and used a chunk size of 1000 and 10 passes. The *alpha* parameter was set to 1.0/number of topics, and the other parameters were set to default settings. The model was run with the number of topics in the following sequence: {5, 10, 15, 20, 25}. In order to select the most appropriate number of topics, the *pyLDAvis gensim* notebook [17] was used to visualize the topics and determine the quality of the model. The *pyLDAvis* notebook visualizes each topic as a bubble, where the area of the bubble resembles the prevalence of the topic in the corpus and inter-topic distances are displayed by means of multidimensional scaling onto two axes [18]. A proper topic model is thus visualized in *pyLDAvis* with relatively big, non-overlapping bubbles. In our case, the most informative number of topics was 10 topics. We will look at the results into detail in the following section.

3   Results

The final model as visualized by *pyLDAvis* is displayed in Fig. 1. Table 1 shows the 15 most relevant terms for each of the 10 topics. The topics are ordered in descending order of prevalence in the corpus. The 15 most salient terms overall are:

cancer, die, health, drug, new, year, old, study, today, yesterday, say, hospital, state, federal, medical

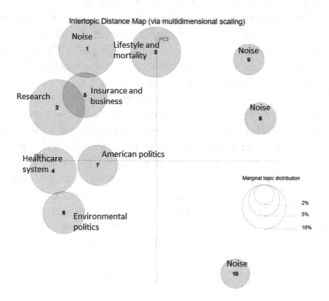

Fig. 1. Final LDA model as visualized by *pyLDAvis,* each bubble represents a topic.

Table 1. Top-15 most relevant terms per topic.

Topic #	Top-15 most relevant terms	Topic content	% of tokens
1	year, day, time, make, life, first, come, good, last, take, get, world, look, man, people	General (noise)	17.2
2	cancer, drug, study, say, find, doctor, new, patient, may, scientist, woman, breast, disease, treatment, report	Research	16.0
3	year, die, old, cancer, new, hospital, york, home, yesterday, former, child, last, lead, smoke, center	Lifestyle and mortality	13.2
4	health, new, state, care, medical, plan, york, say, hospital, would, program, official, center, public, company	Healthcare system	11.1
5	new, city, year, company, york, town, asbestos, rate, jersey, water, high, people, sell, sale, insurance	Insurance and business	10.4
6	today, say, federal, use, chemical, official, case, judge, ban, government, court, agency, state, food, lead	Environmental politics	9.0
7	today, say, president, senate, leader, campaign, bill, senator, vote, republican, would, house, lead, former, week	American politics	8.4
8	yesterday, new, marry, son, award, mrs, richard, daughter, theater, rev, york, robert, music, film, screen	People and performing arts (noise)	5.2
9	man, kill, death, police, evening, brooklyn, son, mother, watch, brother, dead, marijuana, say, daughter, last	Crime (noise)	5.0
10	yesterday, company, share, stock, nuclear, george, smoker, edward, institute, market, retire, air, aim, japan, secretary	General (noise)	4.4

Most of these terms are linked to medical topics, such as the terms 'health', 'drug' and 'medical'. Words like 'state' and 'federal' are also found to be notable, indicating that some documents are politically themed as well. The words 'breast' and 'woman' are also present in the top-30 salient terms overall; no other type of cancer is present in the top-30.

Out of the 10 topics constructed by the model (Table 1), 4 were found to be noisy topics without a distinct link to a subject (topics 1, 8, 9 and 10). In the remaining topics, issues such as research, lifestyle and politics are found. Topic 2 is focused on research on cancer, and contains notable words such as 'study', 'scientist' and 'treatment'. This topic also contains the words 'breast' and 'woman', showing that some of the medical research topic documents appear to report on research on breast cancer.

Topic 3 is focused on the link between lifestyle (such as smoking) and cancer, and people dying because of cancer, with terms such as 'die', 'hospital', 'smoke' and 'cigarette' (top-30). Topic 4 is concerned with the healthcare system with related terms such as 'health', 'plan', 'hospital', 'program' and 'cost' (top-30). Topic 5 addresses topics regarding insurance and business, as it contains terms such as 'company', 'insurance' and 'business' (in the top-30). This topic relates to new technologies developed for cancer treatment by companies and issues with insurance coverage of cancer therapy, hence the more business side of cancer therapy.

Topic 6 and topic 7 are both concerning politics, where the former focuses on environmental politics regarding public safety and the latter focuses on general American politics. The environmental politics topic (topic 6) does not seem to have a clear link with cancer at first sight. However, the terms 'food', 'chemical' and 'pesticide' (top-30) indicate that these documents are linked to news addressing the safety of pesticides in agriculture, their correlation to cancer and legislations for and against pesticides in court. Topic 7 discusses American politics, for which in-depth analysis shows that this topic features mainly issues regarding policies on bills for smoking in public and legalization of marijuana for cancer patients undergoing (chemo) therapy.

The other four topics (1, 8, 9, 10) include noise terms and do not contain a distinct subject, although one topic seems to consist of people and performing arts institutions (topic 8) among its noise and the other crime related elements (topic 9).

4 Discussion and Conclusion

In this section, we discuss the results (Sect. 4.1), provide a more general discussion (Sect. 4.2) and present the conclusion of this paper (Sect. 4.3).

4.1 Discussion of Results

In general, cancer related coverage within the New York Times between 1970 and 2018 seems to center around health and science issues on the one hand, and business and political issues on the other hand. Out of the 10 LDA generated topics, 6 topics contain a distinct subject matter: research, lifestyle and mortality, the healthcare system, business and insurances, environmental politics and American politics. The first three topics are more science and health related while the latter three address business and political issues. The other four topics include noise terms and do not contain a distinct subject.

In particular, the terms 'breast' and 'woman' are salient in the entire corpus, indicating that breast cancer is one of the commonly discussed topics overall. No other organ specific cancer type is mentioned in the top-30 most notable general terms. This result corresponds with previous research, whose results have also shown the predominance of breast cancer in comparison with other types of cancer even though breast cancer did not have the highest incidence rate at that time [1–4]. The larger amount of articles reporting on breast cancer may be due to the highly vocal breast cancer movements [3] such as the Pink ribbon campaign.

In contrast to Clarke and Everest [3], our topics did not show fear or battle metaphors. This may be due to two reasons: (1) Clarke and Everest analyzed magazines rather than newspapers, and the difference in framing of cancer is due to this difference in media type, (2) Clarke and Everest used manual analysis and we have used topic modeling, and the LDA method is not designed for such complex understanding of text beyond solely finding salient words and topics.

Our results do show similarities with the results of Musso and Wakefield [4]. One of our six distinct topics is the healthcare system, which was also a frequently covered subject in the results of Musso and Wakefield [4]. Additionally, both studies have found issues regarding lifestyle (e.g. smoking) and pesticides to be repeatedly addressed in news articles on cancer.

Interestingly, a substantial proportion of the topics found in our research concerns business and political issues regarding cancer. These aspects of cancer were not prevalent in previous studies [1–4]. Hence, this might be a specific area of interest of the New York Times.

4.2 General Discussion

Even though topic modeling has been used for a wide range of applications already, this is the first time that it is applied to science and health communication research. Our paper has made a first step into exploring the usage of topic modeling to this new application area, which is an addition to the large set of manual content analysis techniques currently used in science communication research. Future research could also study the occurrence of topics over time and analyze links to historic events, in order to better understand the coverage of cancer in the New York Times and possible temporal patterns.

In order to answer our research question, we have used topic modeling and LDA in particular. Even though an automatic technique such as topic modeling has benefits in comparison with manual content analysis, as described in the introduction, some downsides also exist. Since topic modeling is a computer model, the model has difficulties with understanding nuances and subtext. Moreover, the automatically created topics ideally resemble categorization of issues or frames based on substantiated theory and would be interpreted that way. This is, however, not guaranteed as the LDA model does not classify based on theory, but on (co-)occurrence of words in documents. Accordingly, testing for statistical, internal and external validity is still difficult to do for LDA and topic models in general [16]. Therefore, it is still up to the researcher to interpret the resulting topics and glean insights and results from them.

Considering that topic models have more difficulty with understanding the meaning of text than humans do, an interesting direction for future research is to incorporate human analysis in the computational process of topic modeling. One or more humans are of course already involved, as the researchers give e.g. 'names' to topics and interpret its content. It would be very interesting to include a larger group of people to help with these tasks, as well as other tasks such as marking important terms for a topic or removing terms from topics that are irrelevant. Additionally, involving more people in the interpretation of the topics decreases the subjectivity of one or two researchers interpreting them alone. Hence, including a larger group of people in topic modeling research could lead to higher quality output. We propose to make the project open for the public to voluntarily engage in, also called 'citizen science', as this would be an opportunity to let the public learn something about (computer) science. Therefore, the second author is planning to set up a project for topic modeling with citizen science in the coming few years, but then for analysis of a different topic.

4.3 Conclusion

The goal of our research was to use topic modeling to investigate the coverage of cancer in news items from the New York Times since 1970. Similar to previous research, our results show a predominance of breast cancer as compared with other types of cancer in the analyzed news articles. Additionally, our topic model shows 6 distinct topics: research on cancer, lifestyle and mortality, the healthcare system, business and insurance issues regarding cancer treatment, environmental politics and American politics on cancer-related policies. Our paper made a first step into using topic modeling in health and science communication research, which is an addition to the manual content analysis techniques currently used in the field. Since topic modeling is a computational technique, the model has more difficulty with understanding the meaning of text than humans do. Therefore, future research will be set up to let the public help with analysis of the topic model.

References

1. Greenberg, R.H., Freimuth, V.S., Bratick, E.A.: A content analytic study of daily newspaper coverage of cancer. Commun. Yearb. **3**(8985), 645–654 (1979)
2. Freimuth, V.S., Greenberg, R.H., DeWitt, J., Romano, R.M.: Covering cancer: newspapers and the public interest. J. Commun. **34**(1), 62–73 (1984)
3. Clarke, J.N., Everest, M.M.: Cancer in the mass print media: fear, uncertainty and the medical model. Soc. Sci. Med. **62**(10), 2591–2600 (2006)
4. Musso, E., Wakefield, S.E.L.: "Tales of mind over cancer": cancer risk and prevention in the canadian print media. Health, Risk Soc. **11**(1), 17–38 (2009)
5. The New York Times Developer Network. https://developer.nytimes.com/. Accessed 28 Aug 2018
6. Lau, J., Collier, N., Baldwin, T.: On–line trend analysis with topic models: #twitter trends detection topic model online. In: Proceedings of COLING 2012: Technical Papers, pp. 1519–1534 (2012)
7. Xie, W., Zhu, F., Jiang, J., Lim, E.P., Wang, K.: TopicSketch: real–time bursty topic detection from Twitter. In: IEEE 13th International Conference on Data Mining, pp. 837–846 (2013)
8. Fang, A., Ounis, I., Habel, P., Macdonald, C., Limsopatham, N.: Topic–centric classification of Twitter user's political orientation. In: Proceedings of the 38th International ACM SIGIR Conference on Research and Development in Information Retrieval, pp. 791–794 (2015)
9. Schwartz, H.A., Eichstaedt, J.C., Kern, M.L., Dziurzynski, L., Ramones, S.M., et al.: Personality, gender, and age in the language of social media: the open-vocabulary approach. PLoS ONE **8**(9), e73791 (2013)
10. Wang, C., Blei, D.M.: Collaborative topic modeling for recommending scientific articles. In: Proceedings of the 17th ACM SIGKDD International Conference on Knowledge Discovery and Data Mining, pp. 448–456 (2011)
11. Jacobi, C., van Atteveldt, W., Welbers, K.: Quantitative analysis of large amounts of journalistic texts using topic modelling. Digital J. **4**(1), 89–106 (2016)
12. Nltk.corpus package. https://www.nltk.org/api/nltk.corpus.html. Accessed 28 Aug 2018
13. Hong, L., Davison, B.: Empirical study of topic modeling in Twitter. In: Proceedings of the First Workshop on Social Media Analytics, pp. 80–88 (2010)

14. Rehurek, R., Sojka, P.: Software framework for topic modelling with large corpora. In: Proceedings of the LREC 2010 Workshop on New Challenges for NLP Frameworks, pp. 45–50 (2010)
15. Spacy. https://spacy.io/. Accessed 28 Aug 2018
16. Blei, D.M., Ng, A.Y., Jordan, M.I.: Latent Dirichlet allocation. J. Mach. Learn. Res. 3(Jan), 993–1022 (2003)
17. PyLDAvis, https://pyldavis.readthedocs.io/. Accessed 28 Aug 2018
18. Sievert, C., Shirley, K.E.: LDAvis: a method for visualizing and interpreting topics. In: Proceedings of the Workshop on Interactive Language Learning, Visualization, and Interfaces, pp. 63–70 (2014)

Model Selection for Multi-directional Ensemble of Regression and Classification Trees

Evgeniya Korneva$^{(\boxtimes)}$ and Hendrik Blockeel

KU Leuven, Leuven, Belgium
{evgeniya.korneva,hendrik.blockeel}@cs.kuleuven.be

Abstract. Multi-directional ensembles of Classification and Regression treeS (MERCS) extend random forests towards multi-directional prediction. The current work discusses different strategies of induction of such a model, which comes down to selecting sets of input and output attributes for each tree in the ensemble. It has been previously shown that employing multi-targets trees as MERCS component models helps reduce both model induction and inference time. In the current work, we present a novel output selection strategy for MERCS component model that takes relatedness between the attributes into account and compare it to the random output selection. We observe that accounting for relatedness between targets has a limited effect on performance and discuss the reasons why it is inherently difficult to improve the overall performance of a multi-directional model by altering target selection strategy for its component models.

Keywords: Versatile models · Ensemble learning ·
Multi-task learning · Multi-directional models

1 Introduction

In practice, data analysis often happens in two steps. First, a task-specific machine learning model is built. Then, this model is used for inference. However, in many cases the user may want to perform different prediction tasks on the same data. A typical example of such a setting is missing data imputation. Moreover, sometimes not all of the prediction tasks are known beforehand. In this case, it appears beneficial to learn a single multi-purpose model of the data set that can be re-used to solve various tasks rather then a number of different special-purpose models. An example of such a versatile model of data is MERCS.

MERCS stands for **M**ulti-directional **E**nsemble of **C**lassification and **R**egression tree**S**. The method is first introduced in [13]. The term *multi-directional* in its name refers to the fact that the resulting model is capable

The research is supported by the Research foundation - Flanders (project G079416N, MERCS).

© Springer Nature Switzerland AG 2019
M. Atzmueller and W. Duivesteijn (Eds.): BNAIC 2018, CCIS 1021, pp. 52–64, 2019.
https://doi.org/10.1007/978-3-030-31978-6_5

of answering any possible query of the data of the form "predict Y from X", as opposed to conventional *uni-directional* models, where the sets of the input and target variables are fixed.

MERCS essentially extends well-known random forests to multi-directional prediction by constructing ensembles of (possibly multi-target) decision trees that have different sets of input and output variables. More specifically, let $\mathcal{RF}(\mathbb{I}, \mathbb{O})$ denote a random forest that predicts a set of output variables $\mathbb{O}$ from a given set of input attributes $\mathbb{I}$. Note that $\mathbb{O}$ can contain more than one attribute. Then, for a given data set D with attributes $\mathbb{A} = \{A_1, A_2, \ldots, A_m\}$, its MERCS model $\mathcal{M}(D)$ is formally defined as:

$$\mathcal{M}(D) = \{\mathcal{RF}_i, \ i = 1 \ldots n | \mathcal{RF}_i = \mathcal{RF}(\mathbb{I}_i, \mathbb{O}_i), \ \mathbb{I}_i \subset \mathbb{A}, \ \mathbb{O}_i \subset \mathbb{A}, \ \mathbb{I}_i \cap \mathbb{O}_i = \emptyset\}$$

One of the key research questions is how to decide *which* trees exactly should be included in the model, i.e., how to determine the sets of input and output attributes, $\mathbb{I}_i$ and $\mathbb{O}_i$ respectively, for every component random forest $\mathcal{RF}_i$ in the ensemble, so that the resulting model allows for prediction in any direction?

As pointed out in [13], this is already possible by learning a set of conventional single-target trees (namely, by learning one random forest per attribute in a data set). Experimental results show, however, that employing multi-target trees as MERCS component models leads to much faster model induction and inference comparing to the straightforward baseline approach. Performance-wise, multi-target tree-based models are often inferior to their baseline counterpart. However, the decrease in predictive accuracy is considered marginal taking into account the gain in speed.

In the experiments conducted in [13], attributes are grouped into target sets randomly, while intuitively it seems that performance of the resulting model can be improved if targets predicted jointly by the same component model are somehow similar, or *related*, because in this case the multi-task model is able to exploit potential dependencies between them [7]. But how to define the notion of *relatedness* and how to determine which attributes in the data set are related and should be therefore predicted together?

The goal of this paper is twofold. First, we apply MERCS to regression task and compare the effect of using multi-target component models to that previously observed in the classification case. Second, we investigate the possibility to improve the predictive performance of the MERCS model consisting of multi-target component models. To that end, we propose a novel MERCS induction strategy that takes relatedness between the attributes into account. We namely discuss two different ways of measuring the relatedness.

The rest of the paper is organized as follows. The next section positions MERCS in the context of multi-task learning and discusses the connection between model selection for MERCS and the problem of identifying related tasks in a multi-task setting. Section 3 formalizes different model selection strategies for MERCS. The results of empirical evaluation of the strategies under consideration are presented in Sect. 4. Final section summarizes the key findings.

2   Related Work

This section discusses the relation between MERCS and multi-task learning.

Multi-task learning (also known as multi-output, multi-objective or multi-target prediction) refers to joint prediction of multiple variables by the same model. A special case of multi-task learning is multi-label classification, when a set of binary labels is assigned to every example.

In the multi-task setting, one often faces the challenge of identifying which tasks can benefit from being learned together, and which should be treated separately. This question appears to be closely related to determining which attributes should be predicted together by the same component model in MERCS.

Breskvar et al. propose random output selection for solving multi-target classification and regression problems, which is extended towards random target grouping strategy for MERCS [2,3]. However, generalizing more sophisticated task grouping approaches proposed in the multi-task learning field towards the multi-directional setting is more challenging.

The biggest difference between MERCS and conventional multi-target models is that the latter are still uni-directional. Indeed, such models can only predict a predefined set of targets and, as opposed to MERCS, are not capable of answering random query of data. In addition, in multi-task learning, one often aims at improving prediction performance of a multi-task model with respect to some predefined principal task, while MERCS is not a task-specific model and is expected to perform reasonably well over all possible tasks.

For instance, Piccart et al. explore inductive transfer in the context of decision tree learning. Pairwise transfer between the targets is measured as the gain in predictive performance that two-target model yields over a single-target one. The authors then propose the algorithm that, given one main target, aims at identifying its best *support set*, that is, the best subset of auxiliary targets that can be added to the model so that the resulting multi-output model shows the best predictive performance with respect to the main target.

Applying this technique to MERCS induction will not help to reduce the size of the model compared to the baseline approach: there will still be km trees in the resulting ensemble. In addition, when the number of attribute in data set is high, computing inductive transfer as defined in [9] for all possible combinations of targets becomes computationally expensive.

An alternative clustering-based approach to discover the relation between binary classification tasks is proposed in [11]. First, empirical measure of mutual relatedness between tasks is estimated. Then, the tasks are clustered into classes of mutually related tasks.

The idea of applying clustering techniques to split data set attributes into disjoint target sets based on their mutual relatedness seems applicable in the MERCS context. However, such a clustering, unless constrained, can result in highly unbalanced clusters with many attributes in some of them and very few in the others. Component models, however, will likely perform poorly if too many targets are needed to be predicted from too few inputs.

3 Selection Strategies for MERCS

Baseline Approach

The most straightforward way to build a versatile tree-based model is to learn a random forest of k single-target trees for every single attribute i in the data set:

$$\mathcal{M}(D) = \mathcal{RF}_i(\mathbb{A} \setminus \{A_i\}, \{A_i\}), \quad i = 1 \ldots m$$

This approach results therefore in the MERCS ensemble of km trees in total. For large m and typical k (e.g., $k = 30$), such a simple baseline strategy leads to quite a large model.

Predicting several targets simultaneously by the same component models is a reasonable solution to that problem.

Random Grouping of Targets

When employing multi-target trees as MERCS component models, one can neglect possible relations between the attributes and group them into sets of targets randomly.

Let us partition the m attributes of a data set into m/p disjoint subsets $\mathbb{A}^j, j = 1, \ldots, m/p$, of p targets each. If an ensemble of k multi-target trees is learned for each subset, the resulting MERCS model contains $n = \frac{km}{p}$ trees, rather than km:

$$\mathcal{M}(D) = \mathcal{RF}_j(\mathbb{A}^j, \mathbb{A} \setminus \mathbb{A}^j), \quad j = 1 \ldots m/p$$

Therefore, even though such an approach does not take into account relatedness between the target attributes, one can certainly expect to observe benefits in terms of speed, since the resulting MERCS model is smaller in size. Similarly to the baseline approach, each attribute is predicted by one random forest, but each random forest simultaneously predicts p targets rather than just one.

Heuristic-Based Grouping of Targets

Suppose we have a way to measure the relatedness between any two attributes A_i and A_j in the data set. Let us denote it as r_{ij}.

We will assign attributes as targets to MERCS component models one by one in a greedy manner. At each step, attribute A_i is assigned to the most related model, i.e., the model that already predicts targets that are highly related to the current attribute. Just as in the random grouping case, the number p of targets predicted by each model (and, therefore, the total number of models n) are specified in advance.

More specifically, we propose assessing the relatedness between the attribute A_i and MERCS component model $\mathcal{RF}_k$ by computing relatedness score defined as follows:

$$\mathcal{R}(A_i, \mathcal{RF}_k) = \begin{cases} 1, & \text{if } |\mathbb{O}_k| = 0, \\ \dfrac{\frac{1}{|\mathbb{O}_k|} \sum\limits_{j: A_j \in \mathbb{O}_k} r_{ij}}{\frac{1}{m} \sum\limits_{j=1}^{m} r_{ij}}, & \text{if } 0 < |\mathbb{O}_k| < p, \\ 0, & \text{if } |\mathbb{O}_k| = p. \end{cases}$$

This definition basically means the following:

- If the model does not have any targets assigned yet, the corresponding relatedness score is set to 1.
- If the model already has some targets assigned (e.g. $0 < |\mathbb{O}_k| < p$), the relatedness score is computed as the ratio of the average relatedness of the current attribute to the model's targets and its average relatedness to all of the attributes in the data set.
  The score will therefore be greater than 1 if model's targets appear more related than attributes on average (and it therefore makes sense to group them), and less than 1 if the opposite is true (and it is therefore better to assign the current attribute to a new or a more related model).
- If the model cannot take any more targets ($|\mathbb{O}_k| = p$), the relatedness score is set to zero.

The final question is how to asses relatedness between any pair of attributes itself (i.e., how to obtain r_{ij}). There are two possible ways in which one can interpret the notion of similarity between the learning tasks.

The first one is to compare the values of the target attributes themselves: if those are somewhat interconnected, the learning tasks are related. Arguably the most straightforward way to assess statistical dependency of two variables is by computing **correlation coefficient**. While expressing the degree of linear dependency between the variables, it can be a reasonable proxy for more complex measures of relation (e.g., mutual information score) since in practice even more complex relationships between two variables often have fairly linear component.

Another way to assess the relatedness between two learning tasks is to see it as a fact that they share the same input features, or *find the same input features important* [4].

The internal estimates made by random forest can be easily used for measuring **feature importance** for predicting a particular target.[1] One can construct a matrix $C = \{c_{ij}\}$, where c_{ij} indicates the importance of the attribute A_j for predicting A_i. The importance scores c_{ij} are obtained by learning a single-target Random Forest for attribute A_i. To speed up the process, we propose learning these random forest on a small random subsample (e.g., 30%) of the data. The correlation coefficient between the rows of the matrix C reflects how similar the corresponding attributes are in terms of what inputs they find important, or, in other words, how much the corresponding attributes are related to each other.

The fundamental difference between the two aforementioned ways of assessing relatedness is that the latter directly takes the needs of the learner (in this case, decision tree) into account, while the former only relies on the tasks themselves.

---

[1] One of the ways of doing this is by calculating *mean decrease impurity* for each attribute. Every node in the trees in the forest corresponds to a binary test on a single attribute, and the locally optimal test is chosen based on the impurity measure. While learning a tree, one can estimate how much each input feature decreases the weighted impurity in a tree. The impurity decrease from each attribute test can be averaged over all trees [1].

Table 1. Data sets used in the experiments

Data set	# instances, n	# attributes, m	Source
andro	48	36	[10]
edm	767	10	[10, 12]
jura	358	18	[6, 10]
oes10	402	314	[10]
oes97	333	279	[10]
rf1	9004	72	[10]
scm20d	8965	77	[10]
slump	102	10	[10, 14]
wq	1059	30	[5]

To sum up, there are four different ways to induce a MERCS model that are the following:

1. Learning single-target component models according to the baseline strategy.
2. Learning multi-target component models grouping attributes determining the target sets randomly.
3. Learning multi-target component models grouping correlated targets together.
4. Learning multi-target component models grouping targets with similar feature importance vectors together.

These strategies are empirically compared in the following section.

4   Experiments

The Data
Table 1 provides a short summary of the data sets used for experiments. The data is a collection of benchmarks data sets for multi-target regression problems. Only numerical attributes (e.g., those with more than 10 distinct values) were considered. The attributes have been scaled to fall in the range between 0 and 1.

Experimental Setup
We compare four different MERCS induction strategies formulated in the previous section in terms of induction and inference time, as well predictive performance[2] of the resulting MERCS models. To that end, 10-fold cross validation is performed.

---

[2] More specifically, the predictive accuracy is evaluated based on the performance on prediction tasks of the following form:

$$A_i \leftarrow \mathbb{A} \setminus \{A_i\}, \quad i = 1 \dots m,$$

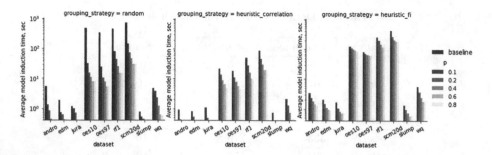

Fig. 1. The more targets are predicted by MERCS component models, the less trees need to be learned. Therefore, model induction becomes faster. However, computing heuristic function to take attribute relatedness into account slows it down.

Besides, when multi-target decision trees are employed, different number of target attributes per component model (parameter p) are considered. More specifically, component models predicting 10%, 20%, 40%, 60% and 80% of the attributes in the data set are considered (p equals to $0.1, 0.2, 0.4, 0.6$ and 0.8 respectively).

Results

Speed. Figures 1 and 2 show average model induction and inference time depending on the number of targets predicted simultaneously by each of the MERCS component models, as well as on the grouping strategy employed.

As expected, the more targets are predicted jointly, the faster the MERCS model can be learned. That is because the total number of trees to be learned decreases. The gain is especially noticeable for the data sets with a large number of attributes. This goes in line with the findings presented in [13]. Grouping targets based on correlation between takes roughly the same time as when grouping them randomly. Indeed, correlation matrix is fast to compute. Obtaining feature importance scores for computing relatedness scores is, by contrast, computationally expensive. One can notice that inducing a MERCS model based on this grouping strategy can actually sometimes take longer than learning single-target component models for each of the attributes. However, if the number of attributes is very high (e.g., data sets oes10, oes97, rf1, scm20d), a speed up in induction time is still observed.

Interesting phenomena can be noticed on the figure below. While inference generally becomes faster when multiple targets are predicted at once by the component models, obtaining predictions is slower when targets are grouped based on their relatedness. That means that the component tree models themselves become larger and more complex.

Performance. To evaluate MERCS predictive performance, we assess the quality of individual attribute prediction by computing the corresponding root mean

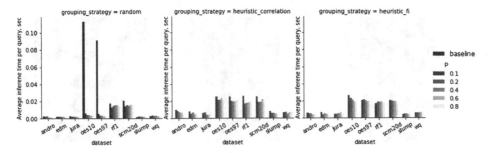

Fig. 2. Inference is generally faster when multiple attributes are predicted by a single component models. Grouping related attributes together results in more complex models, and obtaining predictions takes more time.

squared error (RMSE). We then order the four induction strategies from best to worst for each of the attribute of the data set. The average rank associated with each of the strategy is presented on Fig. 3. Furthermore, Fig. 4 illustrates the distribution of the RMSEs depending on the induction strategy and parameter p for every data set.

To begin with, unlike in the classification case considered in [13], models based on multi-target trees rarely outperform the baseline MERCS model that consists of single-target random forests for each of the attribute. The only example of a data set for which employing multi-target trees in MERCS is actually beneficial is **rf1**. There, the MERCS model with targets grouped randomly outperforms the baseline one for all the values of the parameter p under consideration.

The effect of taking relatedness of the attributes into account when inducing a MERCS model is limited. On the one hand, for most of the data sets, there exist at least one combination of the value of p and heuristic function that on average leads to a more accurate performance with respect to all of the attributes. On the other hand, however, in most cases the difference in the performance appears to be insignificant. For two data sets, namely **edm** and **oes97**, random grouping always works at least as well as more sophisticated heuristic-based approaches.

When comparing the two heuristic-based grouping strategies with each other, one can notice that grouping based on similarity in feature importance scores results in more accurate models for most of the data sets. The only exceptions are data sets **wq** and **edm**: for their attributes, correlation-based grouping is superior.

The next section discusses why employing a more sophisticated grouping strategy does not result in a significant improvement in predictive performance.

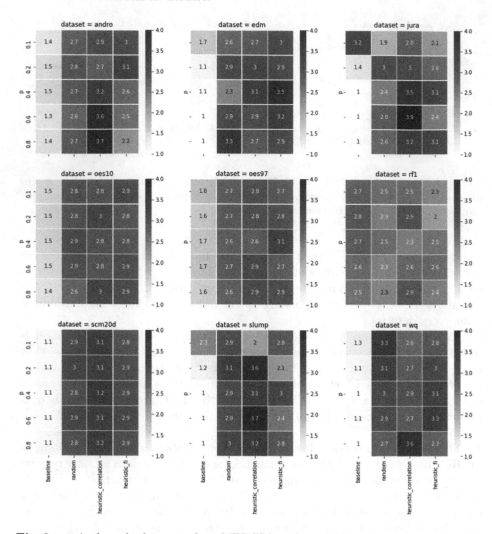

Fig. 3. rf1 is the only data set where MERCS based on multi-target component models outperforms the baseline model for all the number of targets considered. Grouping attributes based on similarity of their feature importance vectors results in more accurate models than when using correlation coefficient as relatedness measure.

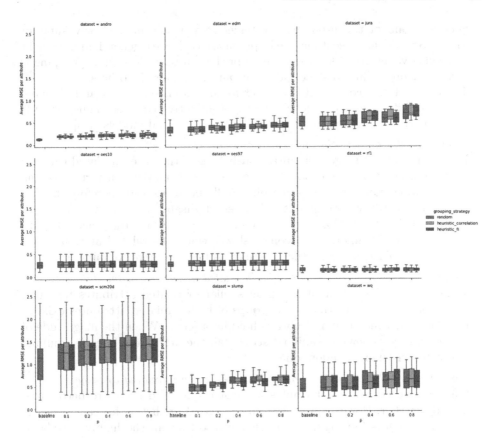

Fig. 4. Employing heuristic to group related attributes together has a limited effect on predictive performance of the resulting MERCS model. For most of the data sets, there is a combination of heuristic function and value of the parameter p that helps improve the performance of the resulting model compared to the random grouping case. However, the improvement appears to be marginal.

5   Discussion

There are a number of possible reasons why grouping attributes into target sets in a 'smart' way does not seem to improve the performance of the resulting MERCS model. They are namely the following.

- *Predicting several targets simultaneously can be beneficial for some of them and detrimental for others.*
  Any combination of the number of outputs p and grouping strategy can result in a MERCS model that is better in solving some prediction tasks and worse in solving others. This observation leads to a number of important remarks. First, average performance will stay roughly the same (this is exactly what has been observed in the results of the experiments conducted in the current paper).

Second, none of the grouping strategies may result in a clearly superior MERCS model because it may be impossible to achieve a general improvement in predictive accuracy for every single prediction task: every new grouping is better for predicting in some directions and worse for the others.

Third, correlation coefficient between two variables is symmetric and therefore does not reflect the asymmetric nature of inductive transfer. Thus, using it as the measure of relatedness will result in suboptimal models [4,9].

- *Models multiple predicting related targets overfit.*
  Predicting several targets simultaneously even if they are unrelated can still be beneficial because additional targets act as regularization terms, forcing the resulting model to perform well on all the tasks and therefore making it less prone to overfitting [4,8]. This effect disappears if targets are very similar. This observation can be confirmed by the fact that, according to our observations, inference becomes slower when the related attributes are grouped together, meaning that the models become more complex.

- *There are possibly no groups of highly related attributes in the data set.*
  When looking for a way of grouping similar or related attributes together, one explicitly assumes that such groups of highly related attributes exist in the data set, which may not always hold in practice. When the magnitude of relatedness is somewhat similar between all the attributes, such a grouping is not better than a random one.

- *Predictors in the component models happen to be unrelated to targets.*
  One can see correlated attributes as related ones and therefore better predicted together. However, they can also be good predictors for each other. If the value of p is too high, it can be the case that all the highly correlated attributes in the data set are grouped together to be predicted by a single component model with he rest of the attributes as predictors. However, since by construction of the model those are less related to the targets, the performance of such a model will be poor.

- *Feature importance scores obtained from a single-target random forest may no longer be reliable when a multi-target one is constructed.*
  Indeed, we group the attributes in a target set for a multi-output component model in such a way that all of them find the same subset of input features important, i.e., splits on these same inputs occur and significantly decrease node impurity when constructing corresponding single-target trees. However, these splits may happen on completely different values. When we then predict the set of targets together, however, the split in a multi-target tree must happen on the same one, and it appears difficult to find a split that would be informative with respect to all of the targets, which results in larger, more complex trees and worse prediction performance.

On balance, employing multi-target random forests helps reduce the size of the final MERCS model in terms of the number of trees that need to be learned. That results in a significant speed-up in model induction, which is the main motivation to use multi-target component models in MERCS. However, there is a trade-off between speed and quality of the predictions: the more targets are predicted jointly, the faster the learning happens, and, typically, the worse the predictions are.

Employing 'smart' grouping strategies may help improve the predictive performance of the multi-target component models. Based on experimental results and aforementioned remarks, we can conclude that grouping attributes based on similarity of the features they find important is more appropriate than relying on the correlation coefficient between the attributes themselves.

However, the optimal value of the parameter p is not known in advance, which is the major shortcoming of the proposed approach. Besides, the resulting MERCS models almost never outperforms the baseline counterpart, which is a MERCS model based on single-target trees. Nevertheless, obtaining feature importance scores is computationally expensive, which can mitigate the gain in learning time if the number of attributes in a data set is not very high.

To sum up, when applying MERCS in practice, one should rather stick to the baseline strategy, especially if the number of attributes in the data set is low. If speed is a priority, opting for random grouping of targets with a relatively low number of targets per component model (e.g., 10% of the total number of attributes) is reasonable, since it can still allow for much faster induction and comparable predictive performance to those in the baseline case. Employing heuristic-based induction strategy with feature importance scores as the measure of relatedness between the attributes can be beneficial for the performance, but extra time is needed to determine the best value of the parameter p.

References

1. Breiman, L.: Random forests. Mach. Learn. **45**(1), 5–32 (2001)
2. Breskvar, M., Kocev, D., Džeroski, S.: Multi-label classification using random label subset selections. In: Yamamoto, A., Kida, T., Uno, T., Kuboyama, T. (eds.) DS 2017. LNCS (LNAI), vol. 10558, pp. 108–115. Springer, Cham (2017). https://doi.org/10.1007/978-3-319-67786-6_8
3. Breskvar, M., Kocev, D., Džeroski, S.: Ensembles for multi-target regression with random output selections. Mach. Learn. **107**(11), 1673–1709 (2018)
4. Caruana, R.: Multitask learning. In: Thrun, S., Pratt, L. (eds.) Learning to Learn, pp. 95–133. Springer, Boston (1998). https://doi.org/10.1007/978-1-4615-5529-2_5
5. Džeroski, S., Demšar, D., Grbović, J.: Predicting chemical parameters of river water quality from bioindicator data. Appl. Intell. **13**(1), 7–17 (2000)
6. Goovaerts, P.: Geostatistics for Natural Resources Evaluation. Oxford University Press on Demand (1997)
7. Kocev, D., Vens, C., Struyf, J., Dzeroski, S.: Tree ensembles for predicting structured outputs. Pattern Recognit. **46**(3), 817–833 (2013). https://doi.org/10.1016/j.patcog.2012.09.023. http://linkinghub.elsevier.com/retrieve/pii/S0031320312300430X
8. Paredes, B.R., Argyriou, A., Berthouze, N., Pontil, M.: Exploiting unrelated tasks in multi-task learning. In: Artificial Intelligence and Statistics, pp. 951–959 (2012)

9. Piccart, B., Struyf, J., Blockeel, H.: Empirical asymmetric selective transfer in multi-objective decision trees. In: Jean-Fran, J.-F., Berthold, M.R., Horváth, T. (eds.) DS 2008. LNCS (LNAI), vol. 5255, pp. 64–75. Springer, Heidelberg (2008). https://doi.org/10.1007/978-3-540-88411-8_9
10. Spyromitros-Xioufis, E., Tsoumakas, G., Groves, W., Vlahavas, I.: Multi-target regression via input space expansion: treating targets as inputs. Mach. Learn. **104**(1), 55–98 (2016). https://doi.org/10.1007/s10994-016-5546-z
11. Thrun, S., O'Sullivan, J.: Discovering structure in multiple learning tasks: the TC algorithm. ICML **96**, 489–497 (1996)
12. Tsanas, A., Xifara, A.: Accurate quantitative estimation of energy performance of residential buildings using statistical machine learning tools. Energy Build. **49**, 560–567 (2012)
13. Van Wolputte, E., Korneva, E., Blockeel, H.: MERCS: multi-directional ensembles of regression and classification trees. In: AAAI Conference on Artificial Intelligence, North America (2018). https://aaai.org/ocs/index.php/AAAI/AAAI18/paper/view/16875
14. Yeh, I.C.: Modeling slump flow of concrete using second-order regressions and artificial neural networks. Cement Concr. Compos. **29**(6), 474–480 (2007)

Finding Dissimilar Explanations in Bayesian Networks: Complexity Results

Johan Kwisthout[(✉)] [iD]

Donders Institute for Brain, Cognition and Behaviour,
Radboud University Nijmegen, PO Box 9104, 6500HE Nijmegen, The Netherlands
j.kwisthout@donders.ru.nl
http://www.socsci.ru.nl/johank

Abstract. Finding the most probable explanation for observed variables in a Bayesian network is a notoriously intractable problem, particularly if there are hidden variables in the network. In this paper we examine the complexity of a related problem, that is, the problem of finding a set of *sufficiently dissimilar*, yet all plausible, explanations. Applications of this problem are, e.g., in search query results (you won't want 10 results that all link to the same website) or in decision support systems. We show that the problem of finding a 'good enough' explanation that differs in structure from the best explanation is at least as hard as finding the best explanation itself.

Keywords: Bayesian networks · MAP explanations · Computational complexity

1 Introduction

A vital computational problem within probabilistic graphical models such as Bayesian networks is the problem of finding the *mode* or *most probable explanation* of a set of variables given observed values for other variables in the network. When the network includes latent or hidden variables (i.e., variables that have neither been observed nor are of interest for the explanation) this problem is known as PARTIAL MAP; the explanation sought is the MAP explanation, i.e., the joint value assignment to the explanation variables that has maximum posterior probability. In this paper we are interested in the problem of finding not specifically the MAP explanation, but *other* explanations that have desirable properties. There are two distinct reasons why we may be interested in alternative explanations:

1. As a means of approximating the MAP explanation. PARTIAL MAP is a highly intractable problem [1,6] and we may find an acceptable approximation thereof (to be further explicated later) 'good enough';

© Springer Nature Switzerland AG 2019
M. Atzmueller and W. Duivesteijn (Eds.): BNAIC 2018, CCIS 1021, pp. 65–72, 2019.
https://doi.org/10.1007/978-3-030-31978-6_6

2. To obtain 'alternative good' explanations *in addition to* the MAP explanation, to allow us to explore several alternatives (e.g. search results) or to cover for a set of likely explanations (e.g. medical conditions).

'Good enough' and 'alternative good' explanations are conceptually different when we look at the *structure* of the explanation as compared to the MAP explanation; that is, how similar or dissimilar the joint value assignments (of the MAP explanation and the alternative explanation) are. In the first case we are more than happy to obtain an explanation that is almost identical to the MAP explanation; in fact, in some problem domains this might even be a prerequisite. For example, in computational cognitive modeling an explanation that does not resemble the MAP explanation would not be acceptable as a valid approximation of the MAP explanation, even if it has a comparable probability. In the second case, our goal is to find alternative explanations that are *structurally different* yet are still plausible. For example, in response to a search query we don't want to end up with ten results that refer to minor variations of essentially the same web-page, even if they happen to be the most probable given the query.

Note that *structure* approximation (where we seek an explanation with Hamming distance at most d of the MAP explanation) is really different from *value* approximation (where we seek an explanation with almost-as-high probability, e.g. within ratio r of the MAP explanation) and from *rank* approximation (where we seek an explanation that is within the m best explanations) [7]. These notions are really orthogonal, as Fig. 1 will reveal.

Current algorithms that seek to find alternative explanations by exploring local maxima (e.g. [2,3,5]) may fare well if the explanatory landscape is as in the "local maxima" panel of Fig. 1. They might not find good explanations, or take a lot of time, if the landscape is in either of the other panels of Fig. 1. However, the computational complexity of this problem has not yet been investigated. In this paper we are specifically interested in explanations that rank well (are within the best m explanations) *and* that are either structurally similar or dissimilar; that is, we complement the results of [7]. In this paper we will further explicate both problems and explore the computational complexity of both of them. We will start with offering some necessary preliminaries and sharing our notational conventions. In Sect. 3 we will formalize both problems (in several variants) and show that both of them are at least as hard as Partial MAP itself. We will also further elaborate on the exact complexity of decision versions of both problems which turns out to be non-trivial. We conclude in Sect. 4.

2   Preliminaries

In this section we introduce our notational conventions. Specifically we will cover Bayesian networks, the complexity classes PP and NP^{PP}, one-Turing reductions, and formal definitions of the approximation notions we will use in the paper. The reader is referred to textbooks like [4] (specifically complexity in Bayesian networks) and to [7] (for a formal treatment of MAP approximations) for more background.

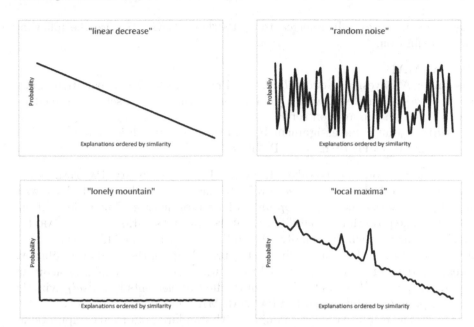

Fig. 1. Graphical depiction of possible relationships between similarity and probability of explanations, with on the X-axis defines an order of the explanations according to their similarity to the best explanation. In "linear decrease" the structure and the probability correlate almost completely; in "random noise" almost not at all. There can be a sole peak of probability mass ("lonely mountain") with all other explanations having almost zero probability. Alternatively, there can be several structurally distinct "local maxima" of probability mass. Particularly in the latter case it might be interesting to look at dissimilar explanations that have a relatively high probability.

A Bayesian network $\mathcal{B} = (\mathbf{G}_\mathcal{B}, \Pr)$ is a probabilistic graphical model that succinctly represents a joint probability distribution $\Pr(\mathbf{V}) = \prod_{i=1}^n \Pr(V_i \mid \pi(V_i))$ over a set of discrete random variables $\mathbf{V}$. $\mathcal{B}$ is defined by a directed acyclic graph $\mathbf{G}_\mathcal{B} = (\mathbf{V}, \mathbf{A})$, where $\mathbf{V}$ represents the stochastic variables and $\mathbf{A}$ models the conditional (in)dependences between them, and a set of parameter probabilities $\Pr$ in the form of conditional probability tables (CPTs). In our notation $\pi(V_i)$ denotes the set of parents of a node V_i in $\mathbf{G}_\mathcal{B}$. We use upper case to indicate variables, lower case to indicate a specific value of a variable, and boldface to indicate sets of variables respectively joint value assignments to such a set.

One of the key computational problems in Bayesian networks is the problem to find the most probable explanation for a set of observations, i.e., a joint value assignment to a designated set of variables (the explanation set) that has maximum posterior probability given the observed variables (the joint value assignment to the evidence set) in the network. If the network includes variables that are neither observed nor to be explained (referred to as intermediate variables) this problem is typically referred to as PARTIAL MAP. We use the following formal definition:

PARTIAL MAP
Instance: A Bayesian network $\mathcal{B} = (\mathbf{G}_\mathcal{B}, \mathrm{Pr})$, where $\mathbf{V}(\mathbf{G}_\mathcal{B})$ is partitioned into a set of evidence nodes $\mathbf{E}$ with a joint value assignment $\mathbf{e}$, a set of intermediate nodes $\mathbf{I}$, and an explanation set $\mathbf{H}$.
Output: A joint value assignment $\mathbf{h}$ to $\mathbf{H}$ such that for all joint value assignments $\mathbf{h}'$ to $\mathbf{H}$, $\mathrm{Pr}(\mathbf{h} \mid \mathbf{e}) \geq \mathrm{Pr}(\mathbf{h}' \mid \mathbf{e})$.

The following notation is taken from [7]. For an arbitrary PARTIAL MAP instance $\{\mathcal{B}, \mathbf{H}, \mathbf{E}, \mathbf{I}, \mathbf{e}\}$, let $cansol_\mathcal{B}$ refer to the set of candidate solutions, with $optsol_\mathcal{B} \in cansol_\mathcal{B}$ denoting the *optimal* solution (or, in case of multiple solutions with the same posterior probability, one of the optimal solutions) to the PARTIAL MAP instance; we will informally refer to this solution as the MAP explanation. When $cansol_\mathcal{B}$ is ordered according to the probability of the candidate solutions (breaking ties between candidate solutions with the same probability arbitrarily), then $optsol_\mathcal{B}^{1\ldots m}$ refers to the set of the first m elements in $cansol_\mathcal{B}$, viz. the m most probable solutions to the PARTIAL MAP instance.

We assume that the reader is familiar with standard notions in computational complexity theory, notably the classes P and NP, NP-hardness, and polynomial time (many-one) reductions. The class PP is the class of decision problems that can be decided by a probabilistic Turing machine in polynomial time; that is, where *Yes*-instances are accepted with probability strictly larger than $1/2$ and *No*-instances are accepted with probability no more than $1/2$. A problem in PP might be accepted with probability $1/2 + \epsilon$ where ϵ may depend exponentially on the input size n. Hence, it may take exponential time to increase the probability of acceptance (by repetition of the computation and taking a majority decision) close to 1. This is consistent with the sampling variant of the Chernoff bound: The number of samples M needed to increase the probability of acceptance of *Yes*-instances to $1 - \delta$ is at least $\frac{\ln(1/\sqrt{\delta})}{\epsilon^2}$; when $\epsilon = 1/2^n$ then M is exponential in the input size. PP hence is a powerful class; we know for example that NP $\subseteq$ PP and the inclusion is assumed to be strict. The canonical PP-complete decision problem is MAJSAT: given a Boolean formula ϕ, does the majority of truth assignments to its variables satisfy ϕ?

In computational complexity theory, so-called *oracles* are theoretical constructs that increase the power of a specific Turing machine. An oracle (e.g., an oracle for PP-complete problems) can be seen as a 'magic sub-routine' that answers class membership queries (e.g, in PP) in a single time step. In this paper we are specifically interested in classes defined by non-deterministic Turing machines with access to a PP-oracle. Such a machine is very powerful, and likewise problems that are complete for the corresponding complexity class NP^{PP} are highly intractable.

A decision variant of PARTIAL MAP is known to be NP^{PP}-complete, even for binary variables, indegree at most 2, and under the assumption that there exists at least one joint value assignment $\mathbf{h}$ such that $\mathrm{Pr}(\mathbf{h}, \mathbf{e}) > 0$ [8,9]. In the intractability proofs in Sect. 3 we will assume, without loss of generality, that these constraints hold for all Bayesian networks $\mathcal{B}$ under consideration. As we use reductions from function problems, not decision problems, our reductions are formally polynomial-time one-Turing reductions. A function f one-Turing reduces to g (notation $f \leq_{1-T}^{\mathsf{FP}} g$) if there are functions t_1 and t_2 such that for all x, $f(x) = t_1(x, g(x, t_2(x)))$ [10, p.5].

We finish this section be repeating the following formal definition of rank-approximation of PARTIAL MAP from [7]; we will build on this definition in the next section.

Definition 1 (rank-approximation of Partial MAP). *Let* $optsol_{\mathcal{B}}^{1\cdots m} \subseteq cansol_{\mathcal{B}}$ *be the set of the m most probable solutions to a* PARTIAL MAP *problem and let* $optsol_{\mathcal{B}}$ *be the optimal solution. An explanation* $approxsol_{\mathcal{B}} \in cansol_{\mathcal{B}}$ *is defined to m-rank-approximate* $optsol_{\mathcal{B}}$ *if* $approxsol_{\mathcal{B}} \in optsol_{\mathcal{B}}^{1\cdots m}$.

3   Main Results

Let d_H be the Hamming distance between two joint value assignments. We define the following two problem variants to PARTIAL MAP, where m and d are arbitrary constants:

(m, d)-SIMILAR PARTIAL MAP
Instance: As in PARTIAL MAP.
Output: An explanation $approxsol_{\mathcal{B}} \in cansol_{\mathcal{B}}$ that m-rank-approximates $optsol_{\mathcal{B}}$ and where $d_H(approxsol_{\mathcal{B}}, optsol_{\mathcal{B}}) \leq d$, or special symbol $\emptyset$ if no such explanation exists.

(m, d)-DISSIMILAR PARTIAL MAP
Instance: As in PARTIAL MAP.
Output: An explanation $approxsol_{\mathcal{B}} \in cansol_{\mathcal{B}}$ that m-rank-approximates $optsol_{\mathcal{B}}$ and where $d_H(approxsol_{\mathcal{B}}, optsol_{\mathcal{B}}) \geq d$, or special symbol $\emptyset$ if no such explanation exists.

We will prove that both problems are $\mathsf{NP^{PP}}$-hard by reduction from PARTIAL MAP, even for binary variables with indegree at most 2. We will start with the construction for (m, d)-SIMILAR PARTIAL MAP without the latter constraints and prove $\mathsf{NP^{PP}}$-hardness (Fig. 2, panel a), and then adapt it to contain only binary variables and indegree at most 2 (panel b). Then we show how this construction can be extended to prove $\mathsf{NP^{PP}}$-hardness of (m, d)-DISSIMILAR PARTIAL MAP (panel c).

Theorem 1 (m, d)-SIMILAR PARTIAL MAP *is* $\mathsf{NP^{PP}}$-*hard.*

Proof. We reduce from the $\mathsf{NP^{PP}}$-hard PARTIAL MAP problem. Let $\{\mathcal{B}, \mathbf{H}, \mathbf{E}, \mathbf{I}, \mathbf{e}\}$ be an instance to PARTIAL MAP. From $\mathcal{B}$, we create an instance $\mathcal{B}'$ to (m, d)-SIMILAR PARTIAL MAP as follows. To $\mathcal{B}$ we add a singleton, unconnected node M with m uniformly distributed values $m_1 \ldots m_m$. Let

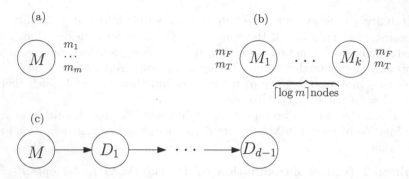

Fig. 2. Nodes added to PARTIAL MAP network $\mathcal{B}$. (a) basic construct; (b) construct with binary nodes only; (c) assuming at least d different nodes.

$\mathbf{H'} = \mathbf{H} \cup \{M\}$ and observe that the MAP explanation $\mathbf{h}$ to $\mathbf{H}$ in the PARTIAL MAP instance now translates into a set of m best explanations $optsol_{\mathcal{B}}^{1...m} = \mathbf{h} \cup \{m_i\}(i = 1 \ldots m)$ with equal probability and by construction all these explanations differ only in the value assignment to M. Let $optsol_{\mathcal{B}}$ be any arbitrary explanation in $optsol_{\mathcal{B}}^{1...m}$ and let apr be a distinct explanation in $optsol_{\mathcal{B}}^{1...m}$. We have that $approxsol_{\mathcal{B}} \in optsol_{\mathcal{B}}^{1...m}$ m-rank-approximates $optsol_{\mathcal{B}}$ and that for any $d \geq 1$, $d_H(approxsol_{\mathcal{B}}, optsol_{\mathcal{B}}) = 1 \leq d$; obviously the one-Turing reduction takes polynomial time and hence (m, d)-SIMILAR PARTIAL MAP is NP^{PP}-hard.

Note that we can replace M in this construction by $k = \lceil \log m \rceil$ uniformly distributed binary variables M_i such that $\mathbf{H'} = \mathbf{H} \cup \{M_1 \ldots M_k\}$. A caveat here is that $\mathbf{h}$ may translate to more than m explanations with equal probability and if $optsol_{\mathcal{B}}^{1...m}$ and $optsol_{\mathcal{B}}$ are picked randomly from this set, we may end up[1] with a set that does not contain $approxsol_{\mathcal{B}}$ such that $d_H(approxsol_{\mathcal{B}}, optsol_{\mathcal{B}}) = 1$. We therefore impose the constraint that $optsol_{\mathcal{B}}^{1...m}$ does not arbitrarily break ties, but that it contains the first m explanations according to their lexicographical order, and because of this we can be sure that $optsol_{\mathcal{B}}^{1...m}$ contains at least one explanation $approxsol_{\mathcal{B}}$ with $d_H(approxsol_{\mathcal{B}}, optsol_{\mathcal{B}}) = 1$.

Corollary 1. (m, d)-SIMILAR PARTIAL MAP *is* NP^{PP}-*hard, even if all nodes are binary and have indegree at most 2.*

We now extend the construction in the proof of Theorem 1 to prove NP^{PP}-hardness of (m, d)-DISSIMILAR PARTIAL MAP. To the above constructed network $\mathcal{B'}$ we add $d - 1$ nodes $D_1 \ldots D_{d-1}$ with m uniformly distributed values. D_1 has M as sole parent, whereas $D_i, i \geq 2$ has D_{i-1} as sole parent. The probability of D_i is defined to be deterministically dependent on its parent, i.e., $\Pr(D_1 = d_{1,j} \mid M = m_j) = 1$ and $\Pr(D_i = d_{i,j} \mid D_{i-1} = d_{i-1,j}) = 1$. We set $\mathbf{H'} = \mathbf{H} \cup \{M\} \cup \{D_1 \ldots D_{d-1}\}$. By construction, every explanation in

---

[1] An example of such a situation would be when $m = 5$ and the solutions with binary encodings 000, 011, 101, 110, and 111 would be in $optsol_{\mathcal{B}}^{1...m}$, with $optsol_{\mathcal{B}} = 000$.

$optsol_{\mathcal{B}}^{1\cdots m}$ will differ in d variables and thus any $approxsol_{\mathcal{B}} \in optsol_{\mathcal{B}}^{1\cdots m}$ m-rank-approximates $optsol_{\mathcal{B}}$ and $d_H(approxsol_{\mathcal{B}}, optsol_{\mathcal{B}}) = d$. The construction with $k = \lceil \log m \rceil$ binary variables M_i is similar, but now we add $(d-1)k$ nodes $D_{1,1} \ldots D_{d-1,k}$, which implies that every $approxsol_{\mathcal{B}} \in optsol_{\mathcal{B}}^{1\cdots m}$ differs in at least d nodes.

Corollary 2. (m,d)-DISSIMILAR PARTIAL MAP *is* $\mathsf{NP^{PP}}$-*hard, even if all nodes are binary and have indegree at most 2.*

Note that m and d are constants and not part of the input. If we would make them part of the input, then unary notation would be necessary as the number of nodes added would otherwise be exponential in the binary representation of m and d. However, since $m \leq 2^n$ (where n denotes the number of variables in $\mathcal{B}$) the reduction would then not be polynomial in the PARTIAL MAP instance any more. Hence, we require m (and d) to be a constant.

3.1   On Membership in $\mathsf{NP^{PP}}$

As explicated before, PARTIAL MAP has an $\mathsf{NP^{PP}}$-complete decision variant [9]. Is it likely that appropriate decision variants of SIMILAR PARTIAL MAP and DISSIMILAR PARTIAL MAP are also in $\mathsf{NP^{PP}}$? We can check in polynomial time, using the MAP explanation $optsol_{\mathcal{B}}$ and the approximation $approxsol_{\mathcal{B}} \in cansol_{\mathcal{B}}$, that the Hamming distance $d_H(approxsol_{\mathcal{B}}, optsol_{\mathcal{B}})$ is as promised. However, a verification algorithm for (m,d)-SIMILAR PARTIAL MAP and (m,d)-DISSIMILAR PARTIAL MAP should also verify that $approxsol_{\mathcal{B}}$ actually is within the m best explanations, and finding the m-th best explanation is known to be $\mathsf{FP^{PP^{PP}}}$-complete [8]. Hence, it is unlikely that (m,d)-SIMILAR PARTIAL MAP and (m,d)-DISSIMILAR PARTIAL MAP have an $\mathsf{NP^{PP}}$-complete decision variant.

4   Conclusion

In this short paper we elaborated on the computational complexity of finding MAP explanations that are almost-as-good as the most probable one (where 'almost-as-good' was defined by rank) *and* that are sufficiently similar or dissimilar to the most probable explanation. We found that these problems are $\mathsf{NP^{PP}}$-hard, that is, not easier than the PARTIAL MAP problem itself. An attempt to extend this proof construct to include value-approximation did not succeed, as the proof constructs really require that all explanations in $approxsol_{\mathcal{B}} \in optsol_{\mathcal{B}}^{1\cdots m}$ have the same probability. In an extreme case it might be that we have just two explanations for a given PARTIAL MAP problem, where $\Pr(h \mid e) = 1/2 + \epsilon$ and $\Pr(\neg h \mid e) = 1/2 - \epsilon$, and any non-uniform probability assignment to the additional variables in the explanation set will destroy the proof.

In future work we'd like to extend our results to value-approximation as well as rank-approximation. It might be relevant to investigate the complexity of these problems in constrained structures, such as polytrees, where the PARTIAL MAP problem nonetheless remains NP-hard.

References

1. de Campos, C.P.: New complexity results for MAP in Bayesian networks. In: Walsh, T. (ed.) Proceedings of IJCAI, vol. 11 (2011)
2. Chen, C., Kolmogorov, V., Zhu, Y., Metaxas, D., Lampert, C.: Computing the M most probable modes of a graphical model. In: Proceedings of the 16th International Conference on Articial Intelligence and Statistics (AISTATS) (2013)
3. Chen, C., Yuan, C., Ye, Z., Chen, C.: Solving M-modes in loopy graphs using tree decompositions. In: Kratochvíl, V., Studený, M. (eds.) Proceedings of the Ninth International Conference on Probabilistic Graphical Models. Proceedings of Machine Learning Research, vol. 72, pp. 145–156 (2018)
4. Darwiche, A.: Modeling and Reasoning with Bayesian Networks. Cambridge University Press, Cambridge (2009)
5. Felzenszwalb, P., Girshick, R., McAllester, D., Ramanan, D.: Object detection with discriminatively trained part-based models. IEEE Trans. Softw. Eng. **32**(9), 1627–1645 (2010)
6. Kwisthout, J.: Most probable explanations in Bayesian networks: complexity and tractability. Int. J. Approx. Reason. **52**(9), 1452–1469 (2011)
7. Kwisthout, J.: Tree-width and the computational complexity of MAP approximations in Bayesian networks. J. Artif. Intell. Res. **53**, 699–720 (2015)
8. Kwisthout, J.H.P., Bodlaender, H.L., van der Gaag, L.C.: The complexity of finding kth most probable explanations in probabilistic networks. In: Černá, I., et al. (eds.) SOFSEM 2011. LNCS, vol. 6543, pp. 356–367. Springer, Heidelberg (2011). https://doi.org/10.1007/978-3-642-18381-2_30
9. Park, J., Darwiche, A.: Complexity results and approximation settings for MAP explanations. J. Artif. Intell. Res. **21**, 101–133 (2004)
10. Toda, S.: Simple characterizations of P(#P) and complete problems. J. Comput. Syst. Sci. **49**, 1–17 (1994)

Beyond Local Nash Equilibria
for Adversarial Networks

Frans A. Oliehoek[1(✉)], Rahul Savani[2], Jose Gallego[3], Elise van der Pol[3],
and Roderich Groß[4]

[1] Delft University of Technology, Delft, The Netherlands
f.a.oliehoek@tudelft.nl
[2] University of Liverpool, Liverpool, UK
[3] University of Amsterdam, Amsterdam, The Netherlands
[4] The University of Sheffield, Sheffield, The Netherlands

Abstract. Save for some special cases, current training methods for
Generative Adversarial Networks (GANs) are at best guaranteed to converge to a 'local Nash equilibrium' (LNE). Such LNEs, however, can be
arbitrarily far from an actual Nash equilibrium (NE), which implies that
there are no guarantees on the quality of the found generator or classifier. This paper proposes to model GANs explicitly as finite games in
mixed strategies, thereby ensuring that every LNE is an NE. We use the
Parallel Nash Memory as a solution method, which is proven to monotonically converge to a resource-bounded Nash equilibrium. We empirically
demonstrate that our method is less prone to typical GAN problems such
as mode collapse and produces solutions that are less exploitable than
those produced by GANs and MGANs.

1 Introduction

Generative Adversarial Networks (GANs) [14] are a framework in which two
neural networks compete with each other: the *generator (G)* tries to trick the
classifier (C) into classifying its generated fake data as true. GANs hold great
promise for the development of accurate generative models for complex distributions without relying on distance metrics [23]. However, GANs are difficult
to train [1,2,40]. A typical problem is *mode collapse*, which can take the form
of *mode omission*, where G does not produce any points from certain modes,
or *mode degeneration*, in which G only partially covers some of the modes. In
fact, except for special cases (cf. Sect. 7), current training methods [17,40] can
only guarantee to converge to a *local* Nash equilibrium (LNE) [35]. However,
an LNE can be arbitrarily far from an NE, and the corresponding generator
might be exploitable by an opponent due to suffering from problems such as
mode collapse. Moreover, adding computational resources alone may not offer
a way to escape these local equilibria: the problem does not lie in the lack of

---

This paper is based on a prior arXiv paper which contains further details [31].

© Springer Nature Switzerland AG 2019
M. Atzmueller and W. Duivesteijn (Eds.): BNAIC 2018, CCIS 1021, pp. 73–89, 2019.
https://doi.org/10.1007/978-3-030-31978-6_7

computational resources, but is inherently the result of only allowing small steps in strategy space using gradient-based training.

We introduce an approach that does not get trapped in LNEs by formulating adversarial networks as *finite* zero-sum games. The solutions that we try to find are saddle points in *mixed strategies*. This approach is motivated by the observation that, in the space of mixed strategies, any LNE is an NE. We employ Parallel Nash Memory (PNM) [29], to search for approximate mixed equilibria with small support.

PNM has been shown to monotonically converge to an NE, provided that in its iterations it has non-zero probability to find better responses [29]. However, due to the extremely large number of pure strategies that result for sensible choices of neural network classes, we cannot expect to find exact best responses. Therefore, we introduce resource-bounded best-responses (RBBRs), and show that our PNM approach monotonically converges to the corresponding resource-bounded Nash equilibrium (RB-NE).

Key features of our approach are that: (1) It is based on finite zero-sum games, and as such it enables the use of existing game-theoretic methods. In this paper we focus on one such method, Parallel Nash Memory (PNM) [29]. (2) It will not get trapped in LNEs: we prove that it monotonically converges to an RB-NE, which means that more computation can improve solution quality. (3) It works for any network architecture. In particular, future improvements in classifiers/generator networks can be exploited directly.

We investigate empirically the effectiveness of PNM and show that it can indeed deal well with typical GAN problems. We show that the found solutions closely match the theoretical predictions made by [14] about the conditions at a Nash equilibrium, and are much less susceptible to being exploited by an adversary than those produced by GANs and MGANs [18].

2   Background

We defer a more detailed treatment of related work until Sect. 7. Here, we introduce some basic game theory.

Definition 1 ('game'). *A two-player strategic game , which we will simply call 'game', is a tuple* $\langle \mathcal{D}, \{\mathcal{S}_i\}_{i \in \mathcal{D}}, \{u_i\}_{i \in \mathcal{D}} \rangle$, *where* $\mathcal{D} = \{1, 2\}$ *is the set of players,* $\mathcal{S}_i$ *is the set of* pure strategies *(actions) for player* i, *and* $u_i : \mathcal{S} \to \mathbb{R}$ *is* $i's$ *payoff function defined on the set of pure strategy profiles* $\mathcal{S} := \mathcal{S}_1 \times \mathcal{S}_2$. *When the action sets are finite, the game is* finite.

We also write s_i and s_{-i} for the strategy of agent i and its opponent respectively. A fundamental concept is the *Nash equilibrium (NE)*, which is a strategy profile $s = \langle s_i, s_{-i} \rangle$ such that no player can unilaterally deviate and improve his payoff: $u_i(s) \geq u_i(\langle s_i', s_{-i} \rangle)$ for all players i and $s_i' \in \mathcal{S}_i$.

A finite game may not possess a pure NE. A *mixed strategy* μ_i of player i is a probability distribution over i's pure strategies $\mathcal{S}_i$. The payoff of a player under a profile of mixed strategies $\mu = \langle \mu_1, \mu_2 \rangle$ is defined as the expectation:

$u_i(\mu) := \sum_{s \in \mathcal{S}} [\prod_{j \in \mathcal{D}} \mu_j(s_j)] \cdot u_i(s)$. Then an NE in mixed strategies is defined as follows. A $\mu = \langle \mu_i, \mu_{-i} \rangle$ is an NE if and only if $u_i(\mu) \geq u_i(\langle s_i', \mu_{-i} \rangle)$ for all players i and potential unilateral deviations $s_i' \in \mathcal{S}_i$. Every finite game has at least one NE in mixed strategies [25]. In this paper we deal with two-player *zero-sum* games, where $u_1(s_1, s_2) = -u_2(s_1, s_2)$ for all $s_1 \in \mathcal{S}_1, s_2 \in \mathcal{S}_2$. The equilibria of zero-sum games, also called *saddle points*,[1] have several important properties, as stated in Von Neuman's Minmax theorem [27]:

Theorem 1. *In a finite zero-sum game, v^* is the* value *of the game that satisfies:* $\min_{\mu_2} \max_{\mu_1} u_1(\mu) = \max_{\mu_1} \min_{\mu_2} u_1(\mu) = v^*$.

All equilibria have payoff v^* and equilibrium strategies are interchangeable: if $\langle \mu_1, \mu_2 \rangle$ and $\langle \mu_1', \mu_2' \rangle$ are equilibria, then so are $\langle \mu_1', \mu_2 \rangle$ and $\langle \mu_1, \mu_2' \rangle$ [32]. This means that in zero-sum games we do not need to worry about equilibrium selection. Moreover, the convex combination of two equilibria is an equilibrium, meaning that the game either has one or infinitely many equilibria. We also employ the standard, additive notion of *approximate equilibrium:* A pair of strategies (μ_i, μ_{-i}) is an ϵ-NE if $\forall i \quad u_i(\mu_i, \mu_{-i}) \geq \max_{\mu_i'} u_i(\mu_i', \mu_{-i}) - \epsilon$.

In the literature, GANs have not typically been considered as finite games. The natural interpretation of the standard setup of GANs is of an infinite game where payoffs are defined over all possible weight parameters for the respective neural networks. With this view we do not obtain existence of saddle points *in the space of parameters*, nor the desirable properties that follow from Theorem 1.[2] This is why the notion of *local Nash equilibrium (LNE)* has arisen in the literature [35,40]. Roughly, an LNE is a strategy profile where neither player can improve in a small neighborhood of the profile. In finite games every LNE is an NE, as, whenever there is a global deviation (i.e., a better response), one can always deviate locally in the space of mixed strategies towards a pure best response (by playing that better response with ϵ higher probability).

3 GANGs

In order to capitalize on the insight that we can escape local equilibria by switching to mixed strategy space for a finite game, we formalize adversarial networks in a finite games setting.[3]

We consider a standard adversarial network setup: $\mathcal{M} = \langle p_d, \langle G, p_z \rangle, C, \phi \rangle$ where

---

[1] Note that in game theory the term 'saddle point' is used to denote a 'global' saddle point which corresponds to a Nash equilibrium: there is no profitable deviation near or far away from the current point. In contrast, in machine learning, the term 'saddle point' typically denotes a 'local' saddle point: no player can improve its payoff by making a small step from the current joint strategy.

[2] Some results on the existence of saddle points in infinite action games are known, but they require properties like convexity and concavity of utility functions [5], which we cannot apply as they would need to hold w.r.t. the neural network parameters.

[3] By relying on Glicksberg's theorem, we think it would be possible to extend our formulation to the continuous setting.

- $p_d(x)$ is the distribution over ('true' or 'real') data points $x \in \mathbb{R}^d$.
- G is a neural network class with d outputs, parametrized by a parameter vector $\theta_G \in \Theta_G$, such that $G(z; \theta_G) \in \mathbb{R}^d$ denotes the ('fake' or 'generated') output of G on a random vector z drawn from some distribution $z \sim p_z$.
- C is a neural network class with a single output, parametrized by a parameter vector $\theta_C \in \Theta_C$, such that the output $C(x; \theta_C) \in [0, 1]$ indicates the 'realness' of x according to C.
- $\phi : [0, 1] \to \mathbb{R}$ is a *measuring function* [4]—e.g., log for GANs, the identity mapping for WGANs—used to specify game payoffs, explained next.

We call $\mathcal{M}$ a *Generative Adversarial Network Game (GANG)*, since it induces a zero-sum game $\langle \mathcal{D} = \{G, C\}, \{\mathcal{S}_G, \mathcal{S}_C\}, \{u_G, u_C\}\rangle$ with:

- $\mathcal{S}_G = \{G(\cdot; \theta_G) \mid \theta_G \in \Theta_G\}$ the set of strategies s_G;
- $\mathcal{S}_C = \{C(\cdot; \theta_C) \mid \theta_C \in \Theta_C\}$ the set of strategies s_C;
- $u_C(s_G, s_C) = \mathbf{E}_{x \sim p_d}[\phi(s_C(x))] - \mathbf{E}_{z \sim p_z}[\phi(s_C(s_G(z)))]$. I.e., the score of C is the expected 'measured realness' of the real data minus that of the fake data;
- $u_G(s_G, s_C) = -u_C(s_G, s_C)$.

As such, when using $\phi = \log$, the above formulation of GANGs employ a payoff function for G that use [14]'s trick to enforce strong gradients early in the training process (but it applies this transformation to u_C too, in order to retain the zero-sum property). It is also possible to use the original GAN objective. Correctness of these transformations is shown in [31].

In practice, GANs are represented using floating point numbers, of which, for a given setup, there is only a finite (albeit large) number. From now on, we will focus on finite GANGs, which have finite parameter sets and a finite set of neural network architectures.

We emphasize this finiteness, because this is exactly what enables us to obtain the desirable properties mentioned in Sect. 2: existence of (one or infinitely many) mixed NEs with the same value, as well as the guarantee that any LNE is an NE. Moreover, these properties hold for the GANG in its original formulation— not for a theoretical abstraction in terms of (infinite capacity) densities—which means that we can truly expect solution methods (that operate in the parametric domain [38]) to exploit these properties. However, since we do not impose any additional constraints or discretization[4], the number of strategies (all possible unique instantiations of the network class with floating point numbers) is *huge*. Therefore, we think that finding (near-) equilibria with small supports is one of the most important challenges for making principled advances in the field of adversarial networks. As a first step towards addressing this challenge, we propose to make use of the *Parallel Nash Memory (PNM)* [29], which can be seen as a generalization (to non-exact best responses) of the *double oracle method* [6, 24].

---

[4] Therefore, our finite GANGs have the same representational capacity as normal GANs that are implemented using floating point arithmetic.

4   Solving GANGs

Treating GANGs as finite games in mixed strategies permits building on existing tools and algorithms for these classes of games [10,11,33]. In this section, we describe how to use Parallel Nash Memory (PNM) [29], which is particularly tailored to find approximate NEs with small support, and monotonically[5] converges to such an equilibrium.

Parallel Nash Memory for GANGs. The basic idea of PNM is that we iteratively find new strategies which are good candidates for improvement of an approximate mixed strategy NE $\langle \mu_G, \mu_C \rangle$. Previous works (such as the original PNM paper [29], and before that the double-oracle method [24]) have considered the use of exact best response (BR) functions to deliver such new candidates. In GANGs, however, computing such an exact BR is intractable, and we typically use gradient descent, or another way to compute an approximate best response. In phrasing our algorithm, we abstract away from the actual implementation of *how* it is computed, but we acknowledge the fact that the quality we can expect is bounded by computational resources. As such we will use the term 'resource-bounded best response' (RBBR) to denote an approximate best response function which computes the best possible answer it can given some amount of computational resources.

Definition 2. *A strategy $s_i \in \mathcal{S}_i^{RB}$ of player i is a* resource-bounded best-response *(RBBR) against a (possibly mixed) strategy μ_j, if*

$$\forall s_i' \in \mathcal{S}_i^{RB}, \quad u_i(s_i, \mu_j) \geq u_i(s_i', \mu_j).$$

That is, s_i only needs to be amongst the best strategies that player i *can compute* in response to μ_j.

For ease of explanation, we focus on the setting with deterministic best responses, but the approach can easily be extended to non-deterministic RBBR functions[6] and our empirical evaluation makes use of such non-deterministic RBBR functions (due to random initializations).

Algorithm 1 details our approach. PNM incrementally grows a strategic game *SG* over a number of iterations using the AUGMENTGAME function. It uses SOLVEGAME to compute (via linear programming, see, e.g., [37]) a mixed strategy NE $\langle \mu_G, \mu_C \rangle$ of *this smaller game* at the end of each iteration. At the beginning of each iteration the algorithm uses the RBBR functions to deliver new

---

[5] For an explanation of the precise meaning of monotonic here, we refer to [29]. Roughly, we will be 'secure' against more strategies of the other agent with each iteration. This does not imply that the worst case payoff for an agent also improves monotonically. The latter property, while desirable, is not possible with an approach that incrementally constructs sub-games of the full game, as considered here: there might always be a part of the game we have not seen yet, but which we might discover in the future that will lead to a very poor worst case payoff for one of the agents.

[6] By changing the termination criterion of line 8 in Algorithm 1 into a criterion for including the newly found strategies. See the formulation in [29] for more details.

Algorithm 1. PARALLEL NASH MEMORY WITH DETERMINISTIC RBBRS

1: $\langle s_G, s_C \rangle \leftarrow$ INITIALSTRATEGIES()
2: $\langle \mu_G, \mu_C \rangle \leftarrow \langle \{s_G\}, \{s_C\} \rangle$ ▷ set initial mixtures
3: **while** True **do**
4:       $s_G \leftarrow$ RBBR(μ_C)   ▷ get new bounded best resp.
5:       $s_C \leftarrow$ RBBR(μ_G)
6:       // Expected payoffs of these 'tests' against mixture:
7:       $u_{BRs} \leftarrow u_G(s_G, \mu_C) + u_C(\mu_G, s_C)$
8:       **if** $u_{BRs} \leq 0$ **then**
9:             **break**
10:      **end if**
11:      $SG \leftarrow$ AUGMENTGAME(SG, s_G, s_C)
12:      $\langle \mu_G, \mu_C \rangle \leftarrow$ SOLVEGAME(SG)
13: **end while**
14: **return** $\langle \mu_G, \mu_C \rangle$ ▷ found an RB-NE

---

promising strategies (s_G, s_C). Then we test if they 'beat' the current $\langle \mu_G, \mu_C \rangle$. If they do, $u_{BRs} > 0$, and the game is augmented with these and solved again to find a new NE of the sub-game SG. If they do not, $u_{BRs} \leq 0$, and the algorithm stops.

AUGMENTGAME evaluates (by simulation) each newly found strategy for each player against all of the existing strategies of the other player, thus constructing a new row and column for the maintained payoff matrix. In order to implement the best response functions, we have used standard stochastic gradient descent, which means that any existing neural network architectures can be used. However, we need to compute RBBRs against *mixtures* of networks of the other player. For C this is trivial: we can simply generate a batch of fake data from the mixture μ_G. Implementing an RBBR for G against μ_C is slightly more involved, as we need to back-propagate the gradient from all the different $s_C \in \mu_C$ to G. Intuitively, one can think of a combined network consisting of the G network with its outputs connected to every $s_C \in \mu_C$. The predictions $\hat{y}_{s_C}$ of these components $s_C \in \mu_C$ are combined in a single linear output node $\hat{y} = \sum_{s_C \in \mu_C} \mu_C(s_C) \cdot \hat{y}_{s_C}$. This allows us to evaluate and backpropagate through the entire network. A practical implementation loops through each component $s_C \in \mu_C$ and does the evaluation of the weighted prediction $\mu_C(s_C) \cdot \hat{y}_{s_C}$ and subsequent backpropagation per component.

Analysis. Given that we do not compute exact BRs, we cannot get convergence to an NE. Instead, using RBBRs, we define an intuitive specialization of NE:

Definition 3. $\mu = \langle \mu_i, \mu_j \rangle$ *is a* resource-bounded NE (RB-NE) *if and only if* $\forall i \; u_i(\mu_i, \mu_j) \geq u_i(RBBR_i(\mu_j), \mu_j)$.

That is, an RB-NE can be thought of as follows: we present μ to each player i and it gets the chance to switch to another strategy, for which it can apply its bounded resources (i.e., use $RBBR_i$) exactly once. After this application, the player's resources are exhausted and if the found $RBBR_i(\mu_j)$ does not lead to a higher payoff it will not have an incentive to deviate.[7]

Intuitively, it is clear that PNM converges to an RB-NE, which we now state formally.

Theorem 2. *If PNM terminates, it has found an RB-NE.*

Proof. We show that $u_{BRs} \leq 0$ implies we have an RB-NE:

$$u_{BRs} = u_G(RBBR_G(\mu_C), \mu_C) + u_C(\mu_G, RBBR_C(\mu_G))$$
$$\leq 0 = u_G(\mu_G, \mu_C) + u_C(\mu_G, \mu_C) \tag{1}$$

Note that, per Definition 2, $u_G(RBBR_G(\mu_C), \mu_C) \geq u_G(s'_G, \mu_C)$ for all computable $s'_G \in \mathcal{S}_G^{RB}$ (and similar for C). Therefore, the only way that $u_G(RBBR_G(\mu_C), \mu_C) \geq u_G(\mu_G, \mu_C)$ could fail to hold, is if μ_G would include some strategies that are not computable (not in $\mathcal{S}_G^{RB}$) that provide higher payoff. However, as the support of μ_G is composed of strategies computed in previous iterations, this cannot be the case. We conclude $u_G(RBBR_G(\mu_C), \mu_C) \geq u_G(\mu_G, \mu_C)$ and similarly $u_C(\mu_G, RBBR_C(\mu_G)) \geq u_C(\mu_G, \mu_C)$. Together with (1) this directly implies $u_G(\mu_G, \mu_C) = u_G(RBBR_G(\mu_C), \mu_C)$ and $u_C(\mu_G, \mu_C) = u_C(\mu_G, RBBR_C(\mu_G))$, indicating we found an RB-NE.

Corollary 1. *Moroever, making use of the finiteness of the game, it can be easily shown that Algorithm 1 terminates and monotonically converges to an equilibrium.*

Proof. This follows directly from the fact that there are only finitely many RBBRs and the fact that we never forget RBBRs that we computed before, thus the proof for PNM [29] extends to Algorithm 1.

Finally, an RB-NE can be linked to the familiar notion of ϵ-NE by making assumptions on the power of the best response computation.

Theorem 3. *If both players are powerful enough to compute ϵ-best responses, then an RB-NE is an ϵ-NE.*

Proof. Starting from the RB-NE (μ_i, μ_j), assume an arbitrary i. By definition of RB-NE $u_i(\mu_i, \mu_j) \geq u_i(RBBR_i(\mu_j), \mu_j) \geq \max_{\mu'_i} u_i(\mu'_i, \mu_j) - \epsilon$.

The PNM algorithm for GANGs is parameter free, but we mention two adaptations that are helpful: Interleaved training of best responses and regularization of classifier best responses. Details can be found in [31].

---

[7] During training the RBBR functions will be used many times. However, the goal of the RB-NE is to provide a characterization of the *end point* of training.

5   Experiments

Here we report on experiments that aim to test if searching in mixed strategies with PNM-GANG can help in reducing problems with training GANs, and if the found solutions (near-RB-NEs) provide better generative models and are potentially closer to true Nash equilibria than those found by GANs (near-LNEs). Since our goal is to produce better generative models, we refrain from evaluating these methods on complex data like images: image quality and log likelihood are not aligned as for instance shown by [39]. Moreover there is debate about whether GANs are overfitting and assessing this from samples is difficult; some methods have been proposed e.g., [3,22,28,36], but most provide merely a measure of variability, not over-fitting. As such, we choose to focus on irrefutable results on mixture of Gaussian (MoG) tasks, for which the distributions can readily be visualized.

Experimental Setup. We compare our PNM approach ('PNM-GANG') to a vanilla GAN implementation and state-of-the-art MGAN [18]. Table 1 summarizes the settings for GAN and PNM training. RBBR models were taken to be as small as possible while still achieving good results. As suggested by [8], we use leaky ReLU as inner activation for our GAN implementation to avoid sparse gradients. Generators have linear output layers. Classifiers use sigmoids for the final layer. Both classifiers and generators are multi-layer perceptrons with 3 hidden layers. We do not use techniques such as Dropout or Batch Normalization, as they did not yield significant improvements in the quality of our experimental results. The MGAN configuration is identical to that of Table 3 in Appendix C1 of [18].

Table 1. Settings used to train GANs and RBBRs.

	GAN	RBBR
Learning rate	$3 \cdot 10^{-4}$	$5 \cdot 10^{-3}$
Batch size	128	128
Dimension of z	40	5
Hidden nodes	50	5
Iterations	20000	750
Generator parameters	4902	92
Classifier parameters	2751	61
Inner activation	Leaky ReLU	Leaky ReLU
Measuring function	log	10^{-5}-bounded log

We test on 3 MoG tasks: 'round', 'grid' and 'random' (cf. Fig. 1). For each we create test cases with 9 and 16 components. In our plots, black points are real data, green points are generated data. Blue indicates areas that are classified as 'realistic' while red indicates a 'fake' classification by C.

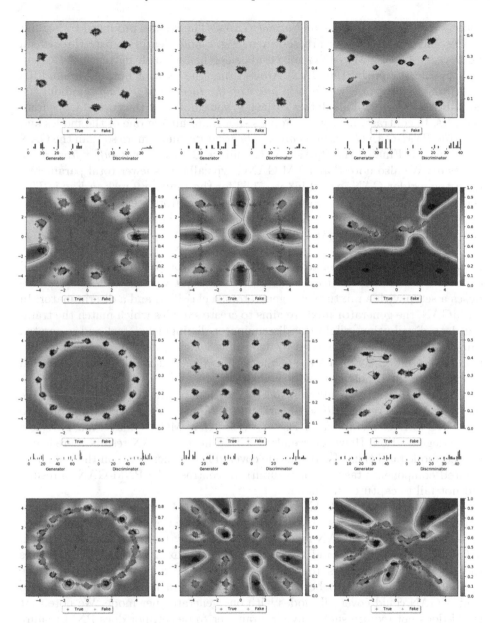

Fig. 1. Results for mixtures of Gaussians with 9 and 16 modes. Odd rows: PNM-GANG, Even rows: GAN. The histograms represent the probabilities in the mixed strategy of each player. True data is shown in black, while fake data is green. The classification boundary (where the classifier outputs 0.5) is indicated with a red line. Best seen in color. (Color figure online)

Found Solutions Compared to Normal GANs. The results produced by regular GANs and PNM-GANGs are shown in Fig. 1 and clearly convey three main points: (1) The PNM-GANG mixed classifier has a much flatter surface than the classifier found by the GAN. Around the true data, it outputs around 0.5 indicating indifference, which is *in line with the theoretical predictions* about the equilibrium [14]. (2) This flatter surface is not coming at the cost of inaccurate samples. In contrast: nearly all samples shown are hitting one of the modes and thus *the PNM-GANG solutions are highly accurate*, much more so than the GAN solutions. (3) Finally, the PNM-GANGs, unlike GANs, *do not suffer from mode omission.* We also note that PNM-GANG typically uses fewer total parameters than the regular GAN, e.g., 1463 vs. 7653 for the random 9 task in Fig. 1. This shows that, qualitatively, the use of multiple generators seems to lead to good results. However, not all modes are *fully* covered. This can be controlled by varying the learning rate [31].

Found Solutions Compared to MGANs. Here we compare the solutions found above for PNM-GANGs to a state-of-the-art GAN variant: MGAN [18] proposes a setup with a mixture of k generators, a classifier, and a discriminator. In an MGAN, the generator mixture aims to create samples which match the training data distribution, while the discriminator distinguishes real and generated samples, and the classifier tries to determine which generator a sample comes from. We use MGAN as a state-of-the art baseline that was explicitly designed to overcome the problem of mode collapse.

Figure 2 shows the results of MGAN on the mixture of Gaussian tasks. We see that MGAN results do seem qualitatively quite good. Comparing them to the PNG-GANG results from Fig. 1, we see that MGAN may even have less mode degeneration. However, we also see that in the MGAN results there is one missed mode (and thus also one mode covered by 2 generators) on the randomly located components task (right column). In contrast, the PNM-GANGs results did not fail to capture any mode.

We point out that MGAN results were obtained with an architecture and hyperparameters which exactly match those proposed by [18] for a similar task. This means that the MGAN models shown use many more parameters (approx. 310,000) than the GAN and GANG models (approx. 2,000). MGAN requires the number of generators to be chosen upfront as a hyperparameter of the method. We chose this to be equal to the number of mixture components, so that MGAN could cover all modes with one generator per mode. We note that PNM does not require such a hyperparameter to be set, nor does PNM require the related "diversity" hyperparameter of the MGAN method (called β in the MGAN paper).

Overall, these results show that the quality of the solutions found by PNM-GANGs is competitive to that of the state-of-the-art MGAN, while using much fewer parameters.

Exploitability of Solutions. Finally, to complement the above qualitative analysis, we also provide a quantitative analysis of the solutions found by GANs, MGANs

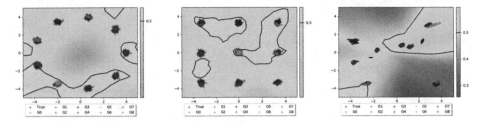

Fig. 2. Results for MGAN on several mixture of Gaussian tasks with 9 modes. Markers correspond to samples created by each generator.

and PNM-GANGs. We investigate to what extent they are exploitable by newly introduced adversaries with some fixed computational power (as modeled by the complexity of the networks we use to attack the found solution). Intuitively, since PNM-GANGs are trained by (against) more powerful attack than GANs, we expect them to be more robust against new attacks of any kind. Specifically, for a given solution $\tilde{\mu} = (\tilde{\mu}_G, \tilde{\mu}_C)$ we use the following measure of exploitability:

$$expl^{RB}(\tilde{\mu}_G, \tilde{\mu}_C) \triangleq \text{RBmax}_{\mu_G} u_G(\mu_G, \tilde{\mu}_C) + \text{RBmax}_{\mu_C} u_C(\tilde{\mu}_G, \mu_C),$$

where 'RBmax' denotes an approximate maximization performed by an adversary of some fixed complexity.

That is, the 'RBmax' functions are analogous to the RBBR functions employed in PNM, but the computational resources of 'RBmax' could be different from those used during PNM. Intuitively, it gives a higher score if $\tilde{\mu}$ is easier to exploit. However, it is not a true measure of distance to an equilibrium: it can return values that are lower than zero which indicate that $\tilde{\mu}$ could not be exploited by the approximate best responses. Our exploitability is closely related to the use of GAN training metrics [20], but additionally includes the exploitability of the classifier. This is important: when only testing the exploitability of the generator, this does give a way to compare generators, *but it does not give a way to assess how far from equilibrium we might be*. Since finite GANGs are zero-sum games, distance to equilibrium is the desired performance measure. In particular, the exploitability of the classifier actually may provide information about the quality of the generator: if the generator holds up well against a perfect classifier, it should be close to the data distribution.[8]

Figure 3 shows our exploitability results for all three tasks with nine modes. We observe roughly the same trend across the three tasks. First, we investigate the exploitability of solutions delivered by GANs, MGANs and GANGs of

---

[8] This measure of exploitability was used to quantify convergence in PNM [29], and also has been used in the optimization literature [26]. It was in the context of GANs in [30], and further motivated for this purpose in [31]. Additionally, an empirical evaluation of exploitability as a measure for GANs was performed in the meantime [16], suggesting that this is a useful measure to quantify sample quality and mode collapse.

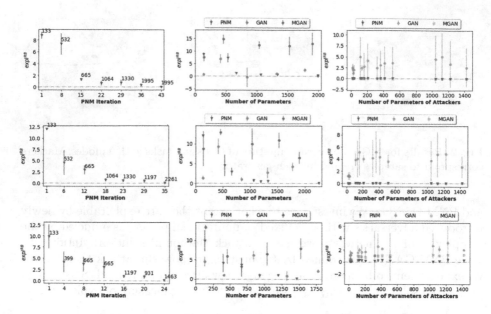

Fig. 3. Exploitability results all 9 mode tasks. Top to bottom: round, grid, random.

different complexities (in terms of *total* number of parameters used). For this, we compute 'attacks' to these solutions using attackers of fixed complexity (a total of 453 parameters for the attacking G and C together). These results are shown in Fig. 3 (left and middle column). The left column shows the exploitability of PNM-GANG after different numbers of iterations, as well as the number of parameters used in the solutions found in those iterations (a sum over all the networks in the support of the mixture). Error bars indicate standard deviation over 15 trials. It is apparent that PNM-GANG solutions with more parameters typically are less exploitable. Also shown is that the variance of exploitability depends heavily on the solution that we happen to attack.

The middle column shows how exploitable GAN, MGAN and PNM-GANG models of different complexities are: the x-axis indicates the total number of parameters, while the y-axis shows the exploitability. The PNM results are the same points also shown in the left column, but repositioned at the appropriate place on the x-axis. All data points are exploitability of models that were trained until convergence. Note that here the x-axis shows the complexity in terms of total parameters. The figure shows an approximately monotonic decrease in exploitability for GANGs, while GANs and MGANs with higher complexity are still very exploitable in many cases. In contrast to GANGs, more complex architectures for GANs or MGANs are thus not necessarily a way to guarantee a better solution.

We also we investigate what happens for the converged GAN/PNM-GANG solution of Fig. 1, which have comparable complexities, when attacked with varying complexity attackers. We also attack the previously reported MGAN solution

(Fig. 2), which has a significantly larger number of parameters (approx. 310,000) than the GAN and GANG models (approx. 2,000). These results are shown in Fig. 3 (right). Clearly shown is that the PNM-GANG is robust with near-zero exploitability even when attacked with high-complexity attackers. The MGAN models also have low exploitability, but recall that these models are *much* more complex. Even with such a complex model, in the 'random' task, the MGAN solution has a non-zero level of exploitability, roughly constant for several attacker complexities. This is most likely related to the missed mode and the fact that two of the MGAN generators collapsed to the same lower-right mode in Fig. 1. In stark contrast to both PNM-GANGs and MGAN, we see that the converged GAN solution is exploitable already for low-complexity attackers, again suggesting that the GAN was stuck in an Local Nash Equilibrium far away from a Nash Equilibrium.

6 Discussion

Overall, the preceding results are very positive: they demonstrate that PNM-trained GANGs can provide more robust solutions than GANs/MGANs with the same number of parameters, suggesting that they are closer to a Nash equilibrium and provide better generative models.

However, we have had (at least so far) less positive results scaling these methods to the image tasks (e.g., MNIST, CelebA or CIFAR-10) that researchers have used to evaluate GANs. A typical problem is that we experience extreme mode collapse of the generator RBBR: i.e., the RBBR typically outputs a single image, regardless of the noise vector z. This mirrors the problem of generator mode collapse also observed in regular GAN training and we are investigating how to overcome this problem by building on techniques, such as minibatch discrimination [36], that were introduced to overcome the related problem in regular GAN training.

An interesting observation that we made on MNIST is that we get the same issues when initializing PNM with the solutions from a number of different runs of normal GAN training. That is, using these GAN solutions as the set of initial strategies, we still find RBBRs (which look like noise attacks) that significantly gain over the mixed strategies maintained, for many iterations. This echos the results that we reported above for the MOG domains: the GAN provided solutions for MNIST are not robust to newly trained attacks. As such, one might question in how far GAN training really works for MNIST in terms of capturing the true data distribution: as far as we can tell these solutions are far from equilibrium, and therefore there is no reason to conclude that the data distribution would be well-captured.

7 Related Work

There is a vast body of work related to this paper. We will restrict to discussing only the most relevant papers here. For a broader discussion, including recent progress on solving in zero-sum games, more general GAN improvements, and bounded rationality, please see [31].

Recently, more researchers have investigated the idea of (more or less) explicitly representing a set or mixture of strategies for the players. For instance, [21] retains sets of networks that are trained by randomly pairing up with a network for the other player thus forming a GAN. This, like PNM, can be interpreted as a coevolutionary approach, but unlike PNM, it does not have any convergence guarantees. MAD-GAN [13] uses k generators, but one discriminator. MGAN [18] proposes mixtures of k generators, a classifier and a discriminator with weight sharing; and presents a theoretical analysis similar to [14] assuming infinite capacity densities. None of these approaches have convergence guarantees.

Generally, explicit mixtures can bring advantages in two ways: *(1) Representation*: intuitively, a mixture of k neural networks could better represent a complex distribution than a single neural network of the same size, and would be roughly on par with a single network that is k times as big. Arora et al. [4] show how to create such a bigger network that is particularly suitable for dealing with multiple modes using a 'multi-way selector'. In our experiments we observed mixtures of simpler networks leading to better performance than a single larger network of the same total complexity (in terms of number of parameters). *(2) Training*: Arora et al. use an architecture that is tailored to representing a mixture of components and train a single such network. We, in contrast, explicitly represent the mixture; given the observation that good solutions will take the form of a mixture. This is a form of domain knowledge that facilitates learning and convergence guarantees.

A closely related paper is the work by [15], which also builds upon game-theoretic tools to give certain convergence guarantees. The main differences are as follows: (1) We provide a more general form of convergence (to an RB-NE) that is applicable to *all* architectures, that only depends on the power to compute best responses, and show that PNM-GANG converges in this sense. We also show that if agents can compute an ϵ-best response, then the procedure converges to an ϵ-NE. (2) [15] show that for a quite specific GAN architecture their first algorithm converges to an ϵ-NE. On the one hand, this result is an instantiation of our more general theory: they assume they can compute exact (for G) and ϵ-approximate (for C) best responses; for such powerful players our Theorem 3 provides that guarantee. On the other hand, their formulation works without discretizing the spaces of strategies. (3) The practical implementation of the algorithm in [15] does not provide guarantees.

Ge et al. [12] propose a method similar to ours that uses *fictitious play* [7,11] rather than PNM. Fictitious play does not explicitly model mixed strategies for the agents, but interprets the opponent's historical behavior as such a mixed strategy. The average strategy played by the 'Fictitious GAN' approach converges to a Nash equilibrium *assuming that "the discriminator and the generator are updated according to the best-response strategy at each iteration"*, which follow from the result by [9] which states that fictitious play converges in continuous zero-sum games. Intuitively, fictitious play, like PNM, in each iteration only ever touches a finite subset of strategies, and one can show that the value of such subgames converges. While this result gives some theoretical underpinning to Fictitious GAN, of course in practice the assumption is hard to satisfy and the notion of RB-NE that we propose may apply to analyze their approach too. Also, in their empirical results they limit the history of actions (played neural networks in previous iterations) to 5 to improve scalability at the cost of convergence guarantees. The Fictitious GAN is not explicitly shown to be more robust than normal GANs, as we show in this paper, but it is demonstrated to produce high quality images, thus showing the potential of game theoretical approaches to GANs to scale.

Hsieh et al. [19] also search in the space of mixed strategies, but without making finiteness assumption (enabled by Glicksberg's theorem). In particular they extend entropic Mirror Descent and Mirror-Prox to infinite dimension to solve GANs, and propose an approximation that can be implemented making use of sampling algorithms. However, for the algorithm to improve over time it needs to make updates that are proportional to the expected payoffs of strategies against the opponents current strategy. De facto this implies that the sampling strategy must be able to find best responses, which in turn implies solving a non-convex optimization problem. As such, it seems unlikely that their practical approach would converge to Nash equilibrium. An interesting question is whether their method can be shown to converge to an RB-NE.

8   Conclusions

We introduce finite GANGs—Generative Adversarial Network Games—a novel framework for representing adversarial networks by formulating them as finite zero-sum games. By tackling them with techniques working in mixed strategies we can avoid getting stuck in local Nash equilibria (LNE). As finite GANGs have extremely large strategy spaces we cannot expect to exactly (or ϵ-approximately) solve them. Therefore, we introduced the resource-bounded Nash equilibrium (RB-NE). This notion is richer than LNE in that it captures not only failures of escaping local optima of gradient descent, but applies to any approximate best-response computations, including methods with random restarts. Additionally, GANGs can draw on a rich set of methods for solving zero-sum games [10, 11, 29, 34]. In this paper, we build on PNM and prove that the resulting method monotonically converges to an RB-NE. We empirically demonstrate that the resulting method does not suffer from typical GAN problems such as mode collapse and forgetting. We also show that the GANG-PNM solutions are closer to theoretical predictions, and are less exploitable than normal GANs: by using PNM we can train models that are more robust than GANs of the same total complexity, indicating they are closer to a Nash equilibrium and yield better generative performance.

We presented a framework that can have many instantiations and modifications. For example, one direction is to employ different learning algorithms.

Another direction could focus on modifications of PNM, such as to allow discarding "stale" pure strategies, which would allow the process to run for longer without being inhibited by the size of the resulting zero-sum "subgame" that must be maintained and repeatedly solved.

Acknowledgments. This research made use of a GPU donated by NVIDIA. F.A.O. is funded by EPSRC First Grant EP/R001227/1. This project had received funding from the European Research Council (ERC) under the European Union's Horizon 2020 research and innovation programme (grant agreement No. 758824—INFLUENCE).

References

1. Arjovsky, M., Bottou, L.: Towards principled methods for training generative adversarial networks. In: ICLR (2017)
2. Arjovsky, M., Chintala, S., Bottou, L.: Wasserstein generative adversarial networks. In: ICML (2017)
3. Arora, S., Zhang, Y.: Do GANs actually learn the distribution? An empirical study, ArXiv e-prints (2017)
4. Arora, S., Ge, R., Liang, Y., Ma, T., Zhang, Y.: Generalization and equilibrium in generative adversarial nets (GANs). In: ICML (2017)
5. Aubin, J.P.: Optima and Equilibria: An Introduction to Nonlinear Analysis, vol. 140. Springer, Heidelberg (1998). https://doi.org/10.1007/978-3-662-03539-9
6. Bosanský, B., Kiekintveld, C., Lisý, V., Pechoucek, M.: An exact double-oracle algorithm for zero-sum extensive-form games with imperfect information. J. AI Res. **51**, 829–866 (2014)
7. Brown, G.W.: Iterative solution of games by fictitious play. Act. Anal. Prod. Alloc. **13**(1), 374–376 (1951)
8. Chintala, S.: How to train a GAN? Tips and tricks to make GANs work. https://github.com/soumith/ganhacks (2016). Accessed 08 Feb 2018
9. Danskin, J.M.: Fictitious play for continuous games revisited. Int. J. Game Theory **10**(3), 147–154 (1981)
10. Foster, D.J., Li, Z., Lykouris, T., Sridharan, K., Tardos, E.: Learning in games: robustness of fast convergence. In: NIPS 29 (2016)
11. Fudenberg, D., Levine, D.K.: The Theory of Learning in Games. MIT Press, Cambridge (1998)
12. Ge, H., Xia, Y., Chen, X., Berry, R., Wu, Y.: Fictitious GAN: training GANs with historical models. ArXiv e-prints (2018)
13. Ghosh, A., Kulharia, V., Namboodiri, V.P., Torr, P.H.S., Dokania, P.K.: Multi-agent diverse generative adversarial networks. ArXiv e-prints (2017)
14. Goodfellow, I., et al.: Generative adversarial nets. In: NIPS 27 (2014)
15. Grnarova, P., Levy, K.Y., Lucchi, A., Hofmann, T., Krause, A.: An online learning approach to generative adversarial networks. In: ICLR (2018)
16. Grnarova, P., Levy, K.Y., Lucchi, A., Perraudin, N., Hofmann, T., Krause, A.: Evaluating GANs via duality. arXiv e-prints (2018)
17. Heusel, M., Ramsauer, H., Unterthiner, T., Nessler, B., Hochreiter, S.: GANs trained by a two time-scale update rule converge to a local Nash equilibrium. In: NIPS 30 (2017)

18. Hoang, Q., Nguyen, T.D., Le, T., Phung, D.Q.: Multi-generator generative adversarial nets. In: ICLR (2018)
19. Hsieh, Y.P., Liu, C., Cevher, V.: Finding mixed Nash equilibria of generative adversarial networks. ArXiv e-prints (2018)
20. Im, D.J., Ma, A.H., Taylor, G.W., Branson, K.: Quantitatively evaluating GANs with divergences proposed for training. In: ICLR (2018)
21. Jiwoong Im, D., Ma, H., Dongjoo Kim, C., Taylor, G.: Generative adversarial parallelization. ArXiv e-prints (2016)
22. Karras, T., Aila, T., Laine, S., Lehtinen, J.: Progressive growing of GANs for improved quality, stability, and variation. In: ICLR (2018)
23. Li, W., Gauci, M., Groß, R.: Turing learning: a metric-free approach to inferring behavior and its application to swarms. Swarm Intell. **10**(3), 211–243 (2016)
24. McMahan, H.B., Gordon, G.J., Blum, A.: Planning in the presence of cost functions controlled by an adversary. In: ICML (2003)
25. Nash, J.F.: Equilibrium points in N-person games. Proc. Natl. Acad. Sci. U. S. A. **36**, 48–49 (1950)
26. Nemirovski, A., Juditsky, A., Lan, G., Shapiro, A.: Robust stochastic approximation approach to stochastic programming. SIAM J. Optim. **19**(4), 1574–1609 (2009)
27. von Neumann, J.: Zur Theorie der Gesellschaftsspiele. Math. Ann. **100**(1), 295–320 (1928)
28. Odena, A., Olah, C., Shlens, J.: Conditional image synthesis with auxiliary classifier GANs. In: ICML (2017)
29. Oliehoek, F.A., de Jong, E.D., Vlassis, N.: The parallel Nash memory for asymmetric games. In: Proceedings of the Genetic and Evolutionary Computation (GECCO) (2006)
30. Oliehoek, F.A., Savani, R., Gallego-Posada, J., Van der Pol, E., De Jong, E.D., Groß, R.: GANGs: generative adversarial network games. ArXiv e-prints (2017)
31. Oliehoek, F.A., Savani, R., Gallego-Posada, J., van der Pol, E., Gross, R.: Beyond local Nash equilibria for adversarial networks. ArXiv e-prints (2018)
32. Osborne, M.J., Rubinstein, A.: Nash equilibrium. In: A Course in Game Theory. The MIT Press (1994)
33. Rakhlin, A., Sridharan, K.: Online learning with predictable sequences. In: COLT (2013)
34. Rakhlin, A., Sridharan, K.: Optimization, learning, and games with predictable sequences. In: NIPS 26 (2013)
35. Ratliff, L.J., Burden, S.A., Sastry, S.S.: Characterization and computation of local Nash equilibria in continuous games. In: Annual Allerton Conference on Communication, Control, and Computing. IEEE (2013)
36. Salimans, T., Goodfellow, I.J., Zaremba, W., Cheung, V., Radford, A., Chen, X.: Improved techniques for training GANs. In: NIPS 29 (2016)
37. Shoham, Y., Leyton-Brown, K.: Multi-Agent Systems: Algorithmic, Game-Theoretic and Logical Foundations. Cambridge University Press, Cambridge (2008)
38. Sinn, M., Rawat, A.: Non-parametric estimation of Jensen-Shannon divergence in generative adversarial network training. In: AISTATS (2018)
39. Theis, L., van den Oord, A., Bethge, M.: A note on the evaluation of generative models. In: ICLR (2016)
40. Unterthiner, T., Nessler, B., Klambauer, G., Heusel, M., Ramsauer, H., Hochreiter, S.: Coulomb GANs: provably optimal Nash equilibria via potential fields. In: ICLR (2018)

Deep Multi-agent Reinforcement Learning in a Homogeneous Open Population

Roxana Rădulescu[1(✉)], Manon Legrand[1], Kyriakos Efthymiadis[1], Diederik M. Roijers[1,2], and Ann Nowé[1]

[1] Vrije Universiteit Brussel, Brussels, Belgium
{roxana.radulescu,kyriakos.efthymiadis,ann.nowe}@vub.be
[2] Vrije Universiteit Amsterdam, Amsterdam, The Netherlands
d.m.roijers@vu.nl

Abstract. Advances in reinforcement learning research have recently produced agents that are competent, or sometimes exceed human performance, in complex tasks. Most interesting real world problems however, are not restricted to one agent, but instead deal with multiple agents acting in the same environment and have proven to be challenging tasks to solve. In this work we present a study on a homogeneous open population of agents modelled as a multi-agent reinforcement learning (MARL) system. We propose a centralised learning approach, with decentralised execution in which agents are given the same policy to execute individually. Using the SimuLane highway traffic simulator as a test-bed we show experimentally that using a single-agent learnt policy to initialise the multi-agent scenario, which we then fine-tune to the task, out-performs agents that learn in the multi-agent setting from scratch. Specifically we contribute an open population MARL configuration, how to transfer knowledge from single- to a multi-agent setting and a training procedure for a homogeneous open population of agents.

Keywords: Multi-agent systems · Reinforcement learning · Open population · Highway traffic

1 Introduction

Recently, a great surge in reinforcement learning research has led to competent artificial agents for increasingly complex tasks, such as playing Atari games and Go [19,25], and robotics [27]. Through these advances, solutions for many real-world problems are now within reach.

A feature of many real-world learning problems is that they require interactions between agents [1,3,7,22]—both human agents, and increasingly also other artificial agents. Such a multi-agent aspect makes these problems

---

M. Legrand—Contribution done during the master thesis studies at the Vrije Universiteit Brussel.

© Springer Nature Switzerland AG 2019
M. Atzmueller and W. Duivesteijn (Eds.): BNAIC 2018, CCIS 1021, pp. 90–105, 2019.
https://doi.org/10.1007/978-3-030-31978-6_8

particularly challenging. Specifically, artificial agents need to learn to antici-pate the behaviour of other agents; typically, failing to do so greatly diminishes the performance. Furthermore, if agents are not fully cooperative (or zero-sum) [15,32], learning algorithms may have trouble identifying a suitable equilibrium policy.

In this paper, we study the behaviour of a population of homogeneous learn-ing agents that is continuously changing, as new agents are entering the system, possibly at different rates than the ones exiting. We characterize this problem as an *open population of homogeneous learning agents*. To tackle this scenario, the agents are sharing the same policy and are learning simultaneously, in the presence of other agents (e.g., human drivers) modelled as part of the environ-ment. The policy-sharing aspect has the advantage that only one policy needs to be learned, and that the agents (both human and artificial) can more easily anticipate the behaviour of the other learning agents. Furthermore, we believe that in many situations humans will be able to more easily predict what the agents will do, which we believe to be a desirable aspect in many systems that also include human agents.

We situate our study in (simulated) traffic on a highway. Not only is this a suitable domain for studying homogeneous agents that share their policy—a car manufacturer probably will want to sell autonomous cars with one policy only—but it also illustrates why we would like to be able to learn around other agents, i.e., humans, and why predictable agents would be desirable. We propose a new simulator for highway traffic, which we briefly discuss in Sect. 3. For a more extensive discussion of the simulator, please refer to [11] and [12].

To learn a policy for our artificial agents, we mainly build upon the Deep Q-learning [18,19] approach. First, we apply this algorithm to learn a policy for a single agent (Sect. 5.1), and test how suitable this policy is for the multi-agent setting. Second, we tackle the multi-agent scenario through the centralized learn-ing decentralized execution paradigm (in which homogeneous learning agents sample simultaneously from an environment to train the same shared neural network that represents their policy) (Sect. 5.2). As expected, this training pro-cedure is significantly slower than single-agent learning. Our key insight is that multi-agent learning can be sped up by first training a single-agent policy, and then using this single-agent policy as a starting point for a homogeneous set of agents. We integrate this key insight into our multi-agent learning algorithm, completing our main contribution. We show experimentally, in Sect. 5.3, that homogeneous multi-agent learning via reuse of a policy trained for a single agent yields better results and is much faster than learning from scratch.

Our contributions for this work can be summarized as follows: (i) we success-fully demonstrate how a single agent policy can be learned in our problem setting and then demonstrate the need for training a model while several autonomous agents are present in the environment, as the single agent one is unable to per-form well in a multi-agent scenario; (ii) we demonstrate two methods for sharing knowledge between agents: either by learning a shared policy using the expe-riences of all the agents in an open population, or by additionally transferring knowledge from a single-agent setting to a multi-agent one, in a complex problem setting.

2 Background

In Reinforcement Learning (RL) [28] an agent learns to solve a task by interacting with the environment, using a numerical reward signal as guidance. Value-based algorithms are a common class of RL techniques. In value-based algorithms, the goal typically is to find an estimation of the action-value function Q defined as the expected sum of rewards discounted at each time step t by a factor γ, when acting according to a policy $\pi = P(a|s)$, defining the probability of any action a in a state s:

$$Q^\pi(s, a) = \mathbb{E}\left[r_t + \gamma r_{t+1} + \gamma^2 r_{t+2} + \cdots \mid s_0 = s, a_0 = a, \pi\right].$$

The optimal value function is then: $Q^*(s, a) = \max_\pi Q^\pi(s, a)$.

Q-learning [31] is a popular value-based RL algorithm, in which the value function is iteratively updated to optimize this expected long-term reward, by bootstrapping the estimated value of the next state. Specifically, after a transition from state s to s', through action a, Q-learning performs the following update:

$$Q(s, a) \leftarrow Q(s, a) + \alpha[r + \gamma \max_{a'} Q(s', a') - Q(s, a)] ,$$

where α is the learning rate, γ is the discount factor and r is the immediate reward received from the environment.

In this paper, we build upon Deep Q-learning, which relies on a Deep Q-network (DQN) [18], i.e., a neural network as a function approximator to estimate the action-value function, $Q(s, a; \theta) \approx Q^*(s, a)$, rather than a tabular representation of the value function. A key aspect for having a stable learning process when introducing this non-linear function approximator is keeping a secondary target network, $Q(s, a; \theta^-)$. In the case of DQN, it suffices to update the target network's parameters every τ steps with the online network [30]. The parameters θ are learned by performing one-step gradient descent updates according to the following loss function at iteration i:

$$L_i(\theta_i) = \mathbb{E}\left(r + \gamma \max_{a'} Q(s', a'; \theta_i^-) - Q(s, a; \theta_i)\right)^2$$

DQN is then used together with *experience replay memory* (ERM) as detailed in [19] to further stabilize and improve the learning process. While an agent interacts with its environment, it gathers experiences of the form $<s_t, a_t, s_{t+1}, r_t>$ at each time step t. In the original Q-learning algorithm [31], each experience is used only once to update the Q-value function. This can be considered wasteful, as certain events may occur with a low frequency. Reusing experiences is the key idea behind the experience replay mechanism [14]; the agent keeps all the past experiences in memory and can then reuse them to update its Q-value function. Additionally, for Deep Q-learning, experience replay plays an important role in breaking the correlation induced by the sequentiality between learning samples.

For the action selection strategy we use ϵ-greedy, choosing a random action with a probability ϵ and the action with the highest Q-value for the rest of the time.

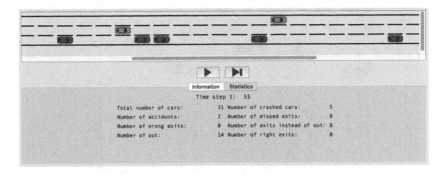

Fig. 1. Graphical Interface of SimuLane. The human drivers are represented in blue, while the autonomous agents are in green. (Color figure online)

3 Problem Setting

In this section, we outline our first contribution: the SimuLane highway traffic simulator.[1]

3.1 Environment

SimuLane is a highway traffic simulator modelling both a single- and multi-agent reinforcement learning environment. It offers a discrete representation of a highway section, where each lane has a preferred speed, reflecting the idea that cars' speed usually increases from right to left on highway lanes. Figure 1 presents a screen-shot of SimuLane's graphical user interface, in which both humans drivers (in blue) and autonomous agents (in green) are interacting in the environment.

Another important component of the simulator are the human drivers. They act according to a behavioural model that can be summarized by three rules: (i) "drivers must not crash", (ii) "drivers must reach their desired speed" and (iii) "drivers must respect the lanes' speeds".

3.2 State and Action Space

Each agent perceives at each time step a state containing only local information, i.e., in accordance to its field of view. The agent is able to observe the presence of other cars, their speed, its own speed and location (i.e., current lane), together with its goal. SimuLane offers learning settings in which the highway can also have exits, with each driver having a certain exit number as a goal. For all the experiments in this work, we do not use exits, thus our agents' goal is to safely traverse the simulated highway segment. The speed of the cars can take integer values in the interval $[0, 3]$. A visual representation of this car-centric input space can be seen in Fig. 2.

---

[1] We have previously demonstrated an earlier version of SimuLane at BNAIC 2017 [12].

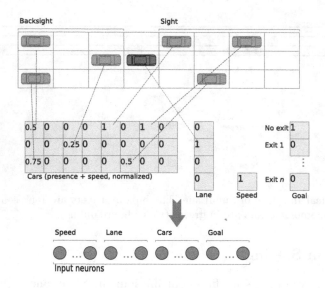

Fig. 2. State space and network inputs for each learning agent in the SimuLane environment. Everything is normalized between $[0, 1]$.

An action is a tuple of the form *<acceleration, direction>*, where the acceleration value is an integer in the range $[-2, 2]$, while the direction can take values from the set {*forward, left, right*}. Our action space is formed by all the possible *<acceleration, direction>* combinations and thus has a size of 15.

3.3 Parameters

Traffic density defines a per lane probability for a car to enter the highway at each time step. The first cell of each lane is thus updated according to the traffic density at every step. For the multi-agent scenario, the *ratio* of autonomous cars is the probability that a new driver entering the highway is an autonomous car.

The size of the highway section can also be configured by setting the *number of lanes* and *cells*. Additionally, the *irrationality* defines the probability for a human driver to choose a random action, allowing one to simulate the fact that a driver can make mistakes.

For the single-agent case we define the maximum number of *steps* the agent can spend on the highway in order to avoid an infinite episode. If the agent is still present on the highway beyond this number of time-steps his outcome will be set to overtime and the episode is terminated. For the multi-agent scenario we fix each episode to a certain number of steps. The *reward function* is also fully configurable for each possible outcome (i.e., crash, overtime and achieved goal).

4 Methods

In this section we outline our second and main contribution: a method for training a homogeneous multi-agent policy in open populations. We are considering here three learning settings: *single-agent*, where only one autonomous car is present on the highway at all times; *multi-agent learning from scratch*, where multiple autonomous cars sharing the same learning model are trained from scratch; and *multi-agent initialized with a single-agent network* as a starting point for the training. We note that for each of the three settings, the human drivers are also present in the environment at all times.

4.1 From Single to Multi-agent Learning

The multi-agent setting is vitally different from the single-agent setting. In a single-agent setting an agent has to learn to respond well to the environment. While this environment may contain other agents, these can generally be assumed to behave according to stationary rules. In our problem setting, we generate such non-learning agents on-the-fly, following a stationary distribution over policies such agents may have. The multi-agent setting on the other hand is considerably harder, i.e., an agent not only has to respond well to the environment, but also to other learning agents – which in our case are running the same, but evolving, policy. This means that the environment is non-stationary, due to these concurrently learning agents. This induces two important challenges: (1) the experience tuples in the experience replay buffer may not resemble the current situation well enough any more, as the learning agent's policies have changed and (2) as a result of this the agents may learn a policy that responds to overly erratic agents, as in the beginning, the learning agents do behave erratically.

Before tackling these problems however, we must first decide upon the basis for the multi-agent approach. As we are considering a homogeneous population of learning agents, we employ Deep Q-learning coupled with a *centralized training with decentralized execution* framework, i.e., during training the agents add their experiences to the same experience replay buffer and only one network is trained for the entire population. At execution time, each agent receives its own network copy and acts according to its own local input. This approach has two main potential advantages: (i) speeding up the training process due to the information sharing between all the agents, (ii) the policy learned will be uniform across the entire population outputting a more predictable behaviour for all the agents. This second point is highly important, as we aim to be able to learn in non-stationary environments.

Our goal for the multi-agent setting is thus to learn one agent-centric policy, uniform over the entire population. As our environment consists in a fixed highway segment through which different agents pass at various rates, we are dealing with an open population. We are learning a policy, by having all agents that pass through the environment contribute, with the experiences they gather, to a central experience replay memory.

For the multi-agent reinforcement learning scenario, as we know from previous research, when employing an experience replay memory [8] learning may become unstable due to non-stationarity. As an initial approach to mitigate this situation we shrink the ERM until we obtained a stable learning process. As an additional helping factor for learning in a multi-agent environment we employ a dropout mask [26] before the output layer. While we expect this approach to mitigate part of the above-stated problems, we do expect the performance of the agents trained in this manner to be significantly worse than in the single-agent case.

4.2   Single to Multi-agent Knowledge Transfer

We identified two problems that may impair learning in a multi-agent reinforcement learning in open populations: the highly erratic behaviour of the agents in the beginning, and the non-stationarity that may render experience replay less useful. Our key insight is that we can significantly reduce the impact of both problems by initializing the policy of the agents with a model learned in the single-agent version of the open-population scenario, i.e., all other agents are non-learning. Firstly, the behaviour of the agents will be significantly better in the beginning, as they already "know" the environment. Furthermore, this will make the behaviour of the learning agents more predictable, and thus make the problem less non-stationary.

Specifically, in our highway traffic problem setting there are two main aspects that have to be learned: (i) driving in a highway environment populated with human drivers, (ii) dealing with the behaviour of other autonomous driving agents. Transfer from the single-agent case to the multi-agent case leverages knowledge about the former from a model trained in a single agent setting, which takes less time to train, and then further tunes this network in order to include in the policy a behaviour adapted for the latter. As we show empirically in Sect. 5, this both reduces the overall training time until convergence with respect to multi-agent network trained from scratch, while the performance is significantly higher.

The fine-tuning procedure we adopt for this work consists in freezing all the weights of the single-agent network model we are transferring to the multi-agent scenario, with the exception of the ones between the last hidden layer and the output. This procedure ensures sufficient exploration in the beginning to escape the risk of running into a local optimum early, while keeping the useful information about the environment encoded in the earlier layers [21]. Additionally, we decrease the learning rate by a factor of 10 in order to stabilize the learning process and avoid unlearning the transferred knowledge.

5   Experiments

In order to test the performance of our algorithms for multi-agent reinforcement learning in open homogeneous populations, we run them on a highway traffic setting, using the settings for our algorithms and our SimuLane simulator stated below. We measure performance of our algorithms as the fraction of times the agents have successfully traversed the highway segment, and thus reached their goal, until each respective episode. We additionally report the fraction of crashes and overtimes.

DQN Settings – We use a DQN function approximator by means of a feed forward network composed of two fully connected layers of size 100 (with a *relu* activation function) and 50 (followed by a *tanh* activation), respectively. There are 42 inputs and 15 outputs. This network architecture is kept the same throughout all our experiments in order to ensure a proper comparison between all the cases (we argue that we look from a policy maker's perspective and aim to learn a car-centred policy in all the settings). The optimizer used is *adagrad* [5] with a learning rate of 0.01. We set two different update intervals for the online and target networks, 10 and 50 respectively, while the batch size for the online network update is set to 16. The experience replay memory has a size of 10 000.

SimuLane Settings – Throughout all our experiments we keep the *traffic density* at a constant 0.25 value, while the highway segment we consider has 3 lanes, each consisting of 40 cells. For the single agent scenario, the learner has 70 time steps to traverse the highway, otherwise the output is set to overtime. For the multi-agent experiments we keep the length of an episode at 160 steps.

Q-learning Settings – The Q-learning parameters are set as follows: ϵ is 0.1, γ is 0.9, while the reward function is set to: goal 1, crash -1, no-speed -0.01, overtime -0.4. The small negative reward for each time step in which the car speed is zero was introduced to encourage agents to avoid the overtime outcome.

5.1   Single-Agent

As a baseline, we first train a policy using only a single agent in the environment, and subsequently test how well that policy performs if it is employed by multiple agents in a multi-agent scenario.

Figure 3 (left) presents the training for the single agent case over 250 000 episodes. The results are averages over 30 runs and we plot the mean and standard deviation for each outcome. We note that for the entire training period the agent keeps a constant ϵ exploration factor. We notice that in training, the agent manages to converge to a performance of over 80%. Additionally, the training curves do not exhibit a high variance, outputting a stable expected performance.

Now that we have a trained network model for the single agent setting, we can test this policy in various scenarios. We begin by looking at how the model behaves in different *traffic densities*. We illustrate the outcomes of our simulations in Fig. 3 (right), under two different traffic density values: 0.25 (the value used during training) and 0.75^2. We average the results over the 30 trained

---

[2] We note that we run simulations for other intermediate values, but only show here the two extremes, for the sake of graph legibility.

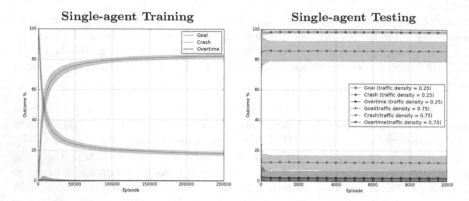

Fig. 3. Single-agent model performance in training (left) and testing (right). The agent keeps a constant ϵ exploration rate of 0.1 for the entire training process. The ϵ is set to 0 during testing. In the testing scenarios, we notice that the further we go away from the value the model was trained under (i.e., 0.25), the bigger the drop in performance gets.

models. One can notice that the further we go away from the value the model was trained under, the bigger the drop in performance will get. This result indicates that, in order to have a better performing model, one should also look at training in a multi-task fashion [6,23], to allow the model to adapt to a bigger variety of environments (at least in terms of traffic density).

A first step towards our goal of having a policy that is able to handle a multi-agent highway scenario is to see how our single agent model performs when multiple learners are present in the environment. We run simulations varying the ratio of autonomous drivers on the highway between 0 and 1, with steps of 0.01. Every agent entering the highway is using the model from the single agent setting in order to act in the environment. The simulations for each ratio last for 500 steps and are repeated 100 times. We plot the mean and standard deviation for each result.

Figure 4 presents the results of these simulations. Notice the downwards trend along the ratio axis in the left subplot, which clearly indicates that a *simple transfer of the single agent policy to a multi-agent setting is not sufficient*. The more autonomous agents are on the highway, the more difficult it is to cope and behave in the environment. On the right side we also take a look at the number of cars entering the highway segment versus the number of cars exiting. We notice also a steep decrease in these values, signalling an increased presence of crashes and stopped cars on the highway, bringing the traffic close to a standstill towards the higher values of the ratio.

5.2   Multi-agent from Scratch

In addition to our single-agent-trained policy as a baseline for multi-agent settings, we also try to train a multi-agent policy from scratch. Please note that

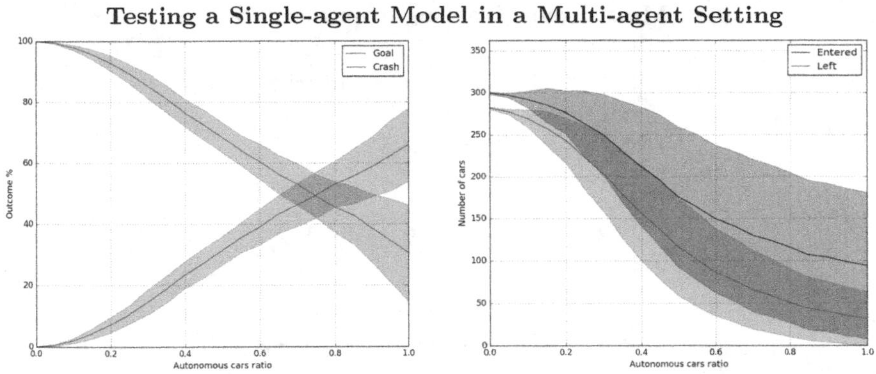

Testing a Single-agent Model in a Multi-agent Setting

Fig. 4. Performance of the single agent model in a multi-agent setting with various ratios of autonomous cars present on the highway. The drop in performance is a clear indication for the need of a model trained in a multi-agent setting, that allows the policy to also incorporate a response to the behaviour of other autonomous agents.

for the multi-agent setting there are two important changes in the parameters mentioned above. The ERM has now a size of 20, and we introduce a dropout mask before the output layer, as explained before, with a value of 0.5. Figure 5 presents the training outcome. The results are averaged over 18 runs and we again run the training process for 250 000 episodes. Notice that in comparison to the single agent case, the expected performance is lower (ending up at a bit over 60% in training), while the variance is higher, showing uncertainty in the final training outcome.

Another important aspect we should note regarding the multi-agent from scratch training procedure is about the training duration. The time for training a multi-agent model from scratch is about 4 times higher than the one for the single agent case (i.e., it took about 4 days, compared to around 20 h).

Moving on to the testing phase, Fig. 6 illustrates the performance of our multi-agent models when various ratios of autonomous cars are present on the highway. Keeping in mind that the ratio under which we trained the policy was 0.5, we can then extract from the left graph the average performance of the models during execution for this scenario, i.e., around 75%. In comparison to Fig. 4, we can definitely notice that the multi-agent models handle better the increase in the autonomous agent ratios, however there is still a higher variance observed towards the second half of the axis. Regarding the number of cars entering the highway, we notice a linear drop, signalling a general decrease in the speed of the cars. Looking also at the number of cars exiting, we notice a slight diverging tendency towards the end, a sign for cars coming to a stand still or getting involved in more crashes.

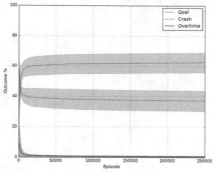

Fig. 5. Performance of the multi-agent from scratch model in training. The agents keeps a constant ϵ exploration rate of 0.1 for the entire process.

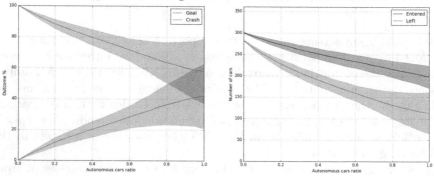

Fig. 6. Performance of the multi-agent model in a setting with various ratios of autonomous cars present on the highway. The drop in performance is not as steep as the single agent case, however there is still much variance in the second half of the graphs.

5.3   Multi-agent with Single-agent Initialization

Finally, we test our main contribution: using a single-agent policy as an initialization to train a multi-agent policy in a homogeneous open population of learners. For this experiment we initialize the network with a model learned in the single agent scenario. We keep the same parameters as in the multi-agent from scratch case, with the exception of the learning rate, which is lowered to 0.001 and the dropout before the output layer, lowered to 0.25. The results are averaged over 8 independent runs.

The results of the fine-tuning process are shown in Fig. 7. We can notice how the goal curve starts now much higher (i.e., at approximately 50%), due to the model initialization, while the final performance level is significantly better compared to the multi-agent learning from zero. The variance is lower and notice

Multi-agent Training with Initialization

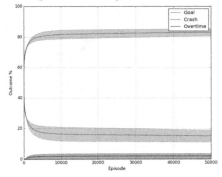

Fig. 7. Performance of the multi-agent with single agent initialization models in training. The agents keeps a constant ϵ exploration rate of 0.1 for the entire process.

how 50 000 additional episodes were already enough to match the single-agent performance. We should also note that the total training time for the single agent model plus the multi-agent initialized from single agent one is still only half of the time required to train the multi-agent network from scratch, while the performance is significantly better.

Our final simulation results (Fig. 8) illustrate how the multi-agent with single agent initialization models perform under various ratios of autonomous cars on the highway. The average performance level for the 0.5 ratio is a bit over 90%. We also notice that the performance has no longer such a steep downwards trend, but always remains above 80%. For the number of cars exiting and entering the highway, even though the variance is fairly high, we can notice the two curves do not diverge from each other, signalling that there is only a decrease in the cars' speed. We can conclude from these results that *models trained in a multi-agent setting, having an initial knowledge transfer from the single agent case exhibit the best performance* in our problem setting.

6   Related Work

Our work takes inspiration from the A3C algorithm [17], in which multiple single-agent environments are run in parallel, and share their experience through pushing gradient updates to a centrally maintained actor and critic. The environment studied in this work however, is vitally different on two important points: in this paper multiple agents act in the *same* environment, and as the population is open the number of agents varies over time.

Our multi-agent training setting is additionally related to the *parameter sharing* training procedure described by [9]. The idea is to leverage the homogeneity of the population of agents and allow them to share the parameters of a common policy. Our open population characteristic, however, prevents us from adding an

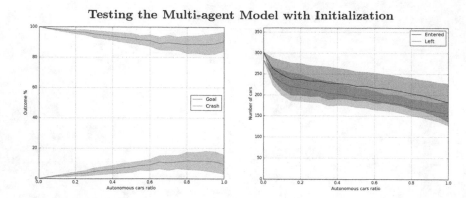

Fig. 8. Performance of multi-agent with single agent initialization models in a setting with various ratios of autonomous cars present on the highway. The drop in performance is only showing a slight decrease, demonstrating that these models are better adapted to a multi-agent scenario.

index for each agent within the model, as the number of agents, as well as the agents themselves, do not remain constant in the population.

In the work of [4] we can also find the idea of an interplay between single and multi-agent scenarios. Although not directly related to our setting, we do find there an interesting procedure for allowing agents to learn as independent learners until there is an explicit requirement for interacting and cooperating with other learners.

The knowledge transfer approach from a single to a multi-agent task presented in this work resembles one of the study cases of [2], where they perform transfer learning from a single-agent predator-prey setting to a multi-agent one (with two or three agents). However, due to our goal of learning a uniform behaviour for each agent in the population, we do not need to establish any mapping between agents for the transfer procedure. We are additionally dealing with a more complex problem setting and a larger number of agents that can enter and exit the system at varying time rates and that are not necessarily performing a cooperative task.

7   Conclusions

In this paper, we looked at the problem of learning in a homogeneous open population of agents, applied in the SimuLane highway traffic simulator. We started by successfully training a single agent in the environment, however that proved to be insufficient in order to cope with a multi-agent setting.

We then examined two basic ways of sharing knowledge between agents: sharing experience from multiple agents in an open population by learning a shared policy, and additionally transferring knowledge from a single-agent setting to a multi-agent setting.

We have shown that in our setting homogeneous multi-agent learning via policy reuse from a single agent yields better results and is much faster than learning from scratch. We thus conclude that transferring from a single-agent policy to a multi-agent policy in homogeneous open-population multi-agent reinforcement learning is both key to keeping learning tractable, and significantly increases performance.

We note that the direct parameter sharing we used for both types of knowledge transfer is only possible due to the homogeneous population. When the set of agents would be heterogeneous, other transfer methods [29] would be required.

In future work, we aim to examine how the learning can be decentralised and still share experience via social learning [10] or by developing a grounded communication system [20]. Additionally, it holds great interest for us to evaluate different (multi-agent) reinforcement learning approaches such as [13,16,24] in this problem setting. Furthermore, another possible study aspect is to remove the homogeneity constraint from the agents and learn in a heterogeneous population. One can then start identifying clusters of similar agents [33] and apply the principle of *learn-from-whom-to-learn* in either a centralized or decentralized manner. Finally, the complexity of the environment can also be increased by considering dynamic traffic densities (simulating daily traffic patterns).

Acknowledgements. This work is supported by Flanders Innovation & Entrepreneurship (VLAIO), SBO project 140047: Stable MultI-agent LEarnIng for neTworks (SMILE-IT), and the European Union FET Proactive Initiative project 64089: Deferred Restructuring of Experience in Autonomous Machines (DREAM) and the Security-Driven Engineering of Cloud-Based Applications (SeCLOUD).

References

1. Amato, C., Oliehoek, F.A.: Scalable planning and learning for multiagent POMDPs. In: AAAI, pp. 1995–2002 (2015)
2. Boutsioukis, G., Partalas, I., Vlahavas, I.: Transfer learning in multi-agent reinforcement learning domains. In: Sanner, S., Hutter, M. (eds.) EWRL 2011. LNCS (LNAI), vol. 7188, pp. 249–260. Springer, Heidelberg (2012). https://doi.org/10.1007/978-3-642-29946-9_25
3. Busoniu, L., Babuska, R., Schutter, B.D.: A comprehensive survey of multiagent reinforcement learning. IEEE Trans Syst. Man Cybern. Part C **38**(2), 156–172 (2008)
4. De Hauwere, Y.M.: Sparse interactions in multi-agent reinforcement learning. Ph.D. thesis, Vrije Universiteit Brussel (2011)
5. Duchi, J., Hazan, E., Singer, Y.: Adaptive subgradient methods for online learning and stochastic optimization. J. Mach. Learn. Res. **12**(Jul), 2121–2159 (2011)
6. Espeholt, L., et al.: IMPALA: scalable distributed deep-RL with importance weighted actor-learner architectures. arXiv preprint arXiv:1802.01561 (2018)
7. Foerster, J., Assael, Y.M., de Freitas, N., Whiteson, S.: Learning to communicate with deep multi-agent reinforcement learning. In: Advances in Neural Information Processing Systems, pp. 2137–2145 (2016)
8. Foerster, J., et al.: Stabilising experience replay for deep multi-agent reinforcement learning. arXiv preprint arXiv:1702.08887 (2017)

9. Gupta, J.K., Egorov, M., Kochenderfer, M.: Cooperative multi-agent control using deep reinforcement learning. In: Sukthankar, G., Rodriguez-Aguilar, J.A. (eds.) AAMAS 2017. LNCS (LNAI), vol. 10642, pp. 66–83. Springer, Cham (2017). https://doi.org/10.1007/978-3-319-71682-4_5

10. Heinerman, J., Rango, M., Eiben, A.E.: Evolution, individual learning, and social learning in a swarm of real robots. In: 2015 IEEE Symposium Series on Computational Intelligence, pp. 1055–1062. IEEE (2015)

11. Legrand, M.: Deep reinforcement learning for autonomous vehicle control among human drivers. Master dissertation, Vrije Universiteit Brussel (2017). http://ai.vub.ac.be/sites/default/files/thesis_legrand.pdf

12. Legrand, M., Rădulescu, R., Roijers, D.M., Nowé, A.: The SimuLane highway traffic simulator for multi-agent reinforcement learning. BNAIC **2017**, 394–395 (2017)

13. Lillicrap, T.P., et al.: Continuous control with deep reinforcement learning. arXiv preprint arXiv:1509.02971 (2015)

14. Lin, L.J.: Self-improving reactive agents based on reinforcement learning, planning and teaching. Mach. Learn. **8**(3–4), 293–321 (1992)

15. Littman, M.L.: Value-function reinforcement learning in Markov games. Cogn. Syst. Res. **2**(1), 55–66 (2001)

16. Lowe, R., Wu, Y., Tamar, A., Harb, J., Abbeel, O.P., Mordatch, I.: Multi-agent actor-critic for mixed cooperative-competitive environments. In: Advances in Neural Information Processing Systems, pp. 6382–6393 (2017)

17. Mnih, V., et al.: Asynchronous methods for deep reinforcement learning. CoRR abs/1602.01783 (2016)

18. Mnih, V., et al.: Playing Atari with deep reinforcement learning. CoRR abs/1312.5602 (2013)

19. Mnih, V., et al.: Human-level control through deep reinforcement learning. Nature **518**(7540), 529–533 (2015)

20. Mordatch, I., Abbeel, P.: Emergence of grounded compositional language in multi-agent populations. arXiv preprint arXiv:1703.04908 (2017)

21. Mossalam, H., Assael, Y., Roijers, D., Whiteson, S.: Multi-objective deep reinforcement learning. In: NIPS Workshop on Deep RL (2016)

22. Nowé, A., Vrancx, P., De Hauwere, Y.M.: Game theory and multi-agent reinforcement learning. In: Wiering, M., van Otterlo, M. (eds.) Reinforcement Learning: State of the Art, pp. 441–470. Springer, Heidelberg (2012). https://doi.org/10.1007/978-3-642-27645-3_14

23. Rusu, A.A., et al.: Progressive neural networks. arXiv preprint arXiv:1606.04671 (2016)

24. Schulman, J., Levine, S., Abbeel, P., Jordan, M., Moritz, P.: Trust region policy optimization. In: International Conference on Machine Learning, pp. 1889–1897 (2015)

25. Silver, D., et al.: Mastering the game of go with deep neural networks and tree search. Nature **529**(7587), 484–489 (2016)

26. Srivastava, N., Hinton, G., Krizhevsky, A., Sutskever, I., Salakhutdinov, R.: Dropout: a simple way to prevent neural networks from overfitting. J. Mach. Learn. Res. **15**, 1929–1958 (2014)

27. Steckelmacher, D., Roijers, D.M., Harutyunyan, A., Vrancx, P., Plisnier, H., Nowé, A.: Reinforcement learning in POMDPs with memoryless options and option-observation initiation sets. AAAI **2018**, 4099–4106 (2018)

28. Sutton, R.S., Barto, A.G.: Reinforcement Learning: An Introduction. MIT Press, Cambridge (1998)

29. Taylor, M.E., Stone, P.: Transfer learning for reinforcement learning domains: a survey. J. Mach. Learn. Res. **10**(Jul), 1633–1685 (2009)
30. Van Hasselt, H., Guez, A., Silver, D.: Deep reinforcement learning with double Q-learning. AAAI **16**, 2094–2100 (2016)
31. Watkins, C.J.C.H.: Learning from delayed rewards. Ph.D. thesis, University of Cambridge England (1989)
32. Wiggers, A.J., Oliehoek, F.A., Roijers, D.M.: Structure in the value function of two-player zero-sum games of incomplete information. In: ECAI 2016, pp. 1628–1629 (2016)
33. Zhang, C., Lesser, V.: Coordinating multi-agent reinforcement learning with limited communication. In: Proceedings of the 2013 International Conference on Autonomous Agents and Multi-agent Systems, pp. 1101–1108 (2013)

Computing and Predicting Winning Hands in the Trick-Taking Game of Klaverjas

Jan N. van Rijn[2,4]($\boxtimes$), Frank W. Takes[3,4], and Jonathan K. Vis[1,4]

[1] Leiden University Medical Center, Leiden, The Netherlands
[2] Columbia University, New York, USA
[3] University of Amsterdam, Amsterdam, The Netherlands
[4] Leiden University, Leiden, The Netherlands

Abstract. This paper deals with the trick-taking game of Klaverjas, in which two teams of two players aim to gather as many high valued cards for their team as possible. We propose an efficient encoding to enumerate possible configurations of the game, such that subsequently $\alpha\beta$-search can be employed to effectively determine whether a given hand of cards is winning. To avoid having to apply the exact approach to all possible game configurations, we introduce a partitioning of hands into 981,541 equivalence classes. In addition, we devise a machine learning approach that, based on a combination of simple features is able to predict with high accuracy whether a hand is winning. This approach essentially mimics humans, who typically decide whether or not to play a dealt hand based on various simple counts of high ranking cards in their hand. By comparing the results of the exact algorithm and the machine learning approach we are able to characterize precisely which instances are difficult to solve for an algorithm, but easy to decide for a human. Results on almost one million game instances show that the exact approach typically solves a game within minutes, whereas a relatively small number of instances require up to several days, traversing a space of several billion game states. Interestingly, it is precisely those instances that are always correctly classified by the machine learning approach. This suggests that a hybrid approach combining both machine learning and exact search may be the solution to a perfect real-time artificial Klaverjas agent.

Keywords: Trick-taking card games · Alpha-beta search · Computational complexity · Machine learning · AI

1 Introduction

A substantial part of artificial intelligence deals with investigating the extent to which machines are able to perform nontrivial complex human tasks. One of such tasks is playing games, a topic which over the years has received a lot of attention

M. Atzmueller and W. Duivesteijn (Eds.): BNAIC 2018, CCIS 1021, pp. 106–120, 2019.
https://doi.org/10.1007/978-3-030-31978-6_9

in artificial intelligence research [8–10], leading to a number of breakthroughs. A recent example is AlphaGo [21], where a combination of search algorithms and machine learning techniques is used to effectively beat humans at the highly complex game of Go. In general, a key problem in such games is that the search space of all possible game configurations is extremely large. This makes it difficult for algorithms to choose for example the next best move, whereas such a decision is often without much effort successfully taken by a human. In this paper we aim to explore this difference in competence of machines and humans, in particular in automatically assessing if a given instance of the game of Klaverjas can be won.

Klaverjas is a trick-taking (card) game, played with the top eight cards from each suit of the French deck. Each card has a face $f \in F = \{7, 8, 9, 10, J, Q, K, A\}$, a suit $s \in S = \{\clubsuit, \diamondsuit, \heartsuit, \spadesuit\}$ and a rank $r \in R = (1, \ldots, 8)$. Cards with higher ranks are considered more powerful. One suit is designated the *trump* suit; cards from this suit are considered more powerful than all cards from other suits. Each card has a specific number of points associated to it, based on the rank and whether it is part of the trump suit or not. Table 1 displays for each card its value in points. Note that the rank of a card depends on the face value and whether it is part of the trump suit.

Table 1. Rank and number of points per card.

Rank	Regular		Trump	
8	A	11	J	20
7	10	10	9	14
6	K	4	A	11
5	Q	3	10	10
4	J	2	K	4
3	9	0	Q	3
2	8	0	8	0
1	7	0	7	0

The game is played with four players that form two teams: one team consists of player N (north) and S (south) whereas the other consists of player E (east) and W (west). Each player starts with eight cards, to be played in each of the eight tricks. A trick is a part of the game in which each player plays one card. The first card of a trick can be freely chosen by the starting player. The other players must follow the leading card. If such a card is not available in a player's hand, a trump card must be played. Whenever a player must play a trump card, and a trump card has already been played in that trick, if possible, a higher rank trump card should be played. If the player can neither follow suit nor play a trump card, it is allowed to play any card that the player has left. It should be noted that there are also versions of the game in which always playing a (higher) trump card is not mandatory if a team mate has already played a trump card, referred to as "Amsterdams" rather than the version which we consider, which is "Rotterdams" Klaverjas.

Once the fourth player has played his card, the trick has ended, and the winner of the trick is determined. If trump cards have been played, the trump card with the highest rank wins the tricks. If not, the card with the highest rank of the leading suit wins the trick. The player who played this card takes all the cards, his team receives the associated points and will start the next trick. The team that wins the last of the eight tricks is awarded 10 additional points. To win the game the team that started the game has to accumulate more points than the opposing team. If they fail to do so, i.e., they lose the game, which is referred to as *nat*, 162 points are awarded to the opposing team. Note that draws do not exist. When the starting team manages to win all eight tricks of the game they are awarded 100 bonus points, referred to as *pit*.

Additionally, special *meld* points can be claimed by the team winning the trick when cards with adjacent face values are in the trick. These are for three ascending face values 20 meld points and for four ascending face values 50 meld points. For the King and Queen of trump, players can claim 20 meld points, in addition to other meld points already claimed in that trick. Finally, when four cards of the same face are played in the same trick, the team winning the trick can claim 100 meld points. For determining meld points, the order in which the players have played these cards is irrelevant. The addition of meld changes the dynamics of the game drastically, as players might sometimes be inclined to play a good card in a trick that is already lost, to prevent conceding meld points to the opposing team. Teams can choose not to claim meld points, for example when they already know they will lose the game.

In this paper we consider the task of determining and predicting whether an instance of the game of Klaverjas is winning for a variant of the game, in which complete information on a team's cards is available. In addition, we assume that there is no bidding process of determining which player starts the game; the first player always starts and determines the trump suit. For this simplified version of the game, we consider the decision problem of, given a particular distribution of cards over players, determining whether this hand is winning for the starting player. We do so using both an exact algorithm based on $\alpha\beta$-search, as well as using a machine learning algorithm that based on feature construction mimics how a human decides whether a hand would be winning, for example based on counting high value cards.

The results presented in this paper are useful for at least two types of new insights. First, in the real game, determining the starting player is done based on bidding, where players assess the quality of their dealt hand, based on whether they think they can win that hand. The approaches presented in this paper essentially perform this type of hand quality assessment. Second, as we employ both an exact approach and a machine learning approach, we can investigate the extent to which both are able to efficiently solve the game of Klaverjas. This will allow to investigate whether exact algorithms have the same difficulties with certain hands as an exact algorithm faces.

The remainder of this paper is organized as follows. After discussing related work in Sect. 2, we introduce various definitions and necessary notation in Sect. 3. Then, an exact algorithm for solving a game of Klaverjas is presented in Sect. 4. Next, a machine learning approach is presented in Sect. 5. A comparison between the two is made in Sect. 6. Finally, Sect. 7 concludes the paper and provides suggestions for future work.

2   Related Work

Klaverjas is an example of the Jack-Nine card games, which are characterized as *trick-taking* games where the Jack and nine of the trump suit are the highest-ranking trumps, and the tens and aces of other suits are the most valuable cards of these suits [16]. Trick-taking games are games of finite length, where players have a hand of cards, and in each round (called a trick) all players play a card from their hand; the player that played the best card according to the rules wins the trick. As Jack-Nine games are not extensively studied in literature, we review some seminal research on solving games, as well as relevant literature on trick-taking games in general.

Exhaustive search strategies have extensively been applied in a number of games [9,19]. For two-player games, the minimax algorithm and its extension $\alpha\beta$-pruning, together henceforth referred to as $\alpha\beta$-search, traverse a game tree to evaluate a given game position. In practice, this is often combined with a *heuristic* evaluation function, in case the game tree is too massive to traverse to all relevant leafs. Historically, much research has been conducted to handcraft *static* heuristic functions that capture human knowledge about the game. Alternative to using a static heuristic function, induction techniques can be used to learn such functions based on previous games. Recently, this strategy has been successfully employed within AlphaGo, an artificial intelligence agent that has beaten the human world champion at Go [21,22]. When no heuristic function is used, and the relevant part of the game tree is completely traversed, the minimax algorithm results in the game-theoretical value, i.e., the outcome of the game assuming perfect play, of this game state [12,17]. This is one of the principal aims of this work.

When agents are confronted with imperfect information, Perfect Information Monte Carlo (PIMC) search is a practical technique for playing games that are too large to be optimally solved [7,15]. PIMC builds a game tree starting with a probabilistic node, branching to all possible configurations. For each configuration, it assumes perfect information and uses common search algorithms, such as minimax with $\alpha\beta$-pruning. It has been noted that the correct card to play might be based on information that the player can not possibly know [6].

Several computational complexity results have been obtained for trick-taking games where all players have perfect information [2,23,24]. According to [2], the natural decision question for trick-taking games is whether a given team can obtain a given number of tricks. In order to obtain computational complexity results, generalizations need to be made. In the case of trick-taking games this can be done over the number of suits, the number of cards per suit, the number of players and the way these players are assigned to two teams. In [23] it was shown that trick-taking games with two players and one suit are in P. In [24], it was proven that trick-taking games with two players, where both players have for each suit an equal number of cards, are in P, for an unbounded number of suits. Later, the authors of [2] showed that all generalizations are in PSPACE, and when generalizing over the number of teams and the number of suits, trick-taking games are PSPACE-complete. Furthermore, they showed that a game with six players and unbounded number of suits and cards per suit is PSPACE-complete. It is shown that the obtained complexity results also apply to trick-taking games that feature a trump suit, such as the Nine-Jack games. The biggest difference between the obtained complexity results and the main interest of our work, Klaverjas, is that the natural decision question of Klaverjas involves a threshold on the number of points, rather than the number of obtained tricks. This decision question requires a different approach.

Other related research regarding trick-taking games focuses on combining and applying search and heuristic evaluation techniques to Skat, a related trick-taking game. We review some examples. In [13], an agent is proposed that uses exact search, assuming perfect information. The implementation features a specific move ordering, transposition tables and adversarial heuristics to speed up the procedure. The main claim is that a typical instance of Skat (with perfect information) can be solved within milliseconds. This conclusion is in line with our (Klaverjas) experiments in Sect. 6.2. The outcomes of individual games can be combined using Monte Carlo techniques. In [5] an agent is proposed that applies inference techniques to both heuristic evaluations and the bidding in Skat. Finally, in [14] a post processing opponent modelling technique is introduced, determining the skill level of an opponent based on earlier games. This is based on the premise that when playing against weaker opponents, moves that have a higher risk and reward pay-off can be played. All mentioned agents use a variant of PIMC to traverse the search space.

3   Preliminaries

We allow ourselves to build upon the excellent definition given by the authors of [2]. The game is played with 32 cards, each player obtaining 8 cards. For each card c, $face(c)$ denotes the face, $suit(c)$ denotes the suit, $rank(c)$ denotes the rank, and $points(c)$ denotes the number of points associated with this card. Note that each card is defined by its suit and face value. The rank and the score follow from this.

A position p is defined by a tuple of hands $h = (h_N, h_E, h_S, h_W)$, where a hand is a set of cards, a trump suit $\phi \in \{\clubsuit, \diamondsuit, \heartsuit, \spadesuit\}$ and a *lead player* $\tau \in P$, where $P = \{N, E, S, W\}$. Let h_p with $p \in P$ be the set of all cards (i.e., the hand) of player p. All players have an equal number of cards, i.e., for all $i, j \in P$

it holds that $|h_i| = |h_j|$. Furthermore, the hands of players do not overlap, i.e., for all $i, j \in P$ (with $i \neq j$) we have $h_i \cap h_j = \emptyset$.

We define $h_{p,s}$ with $p \in P$ be the set of all cards of player p that are of suit s, i.e., $H_{p,s} = \{c \in H_p : suit(c) = s\}$. Let o_p with $p \in P$ be the set of all cards in the opposing team of player p, i.e., $o_N = o_S = h_W \cup h_E$ and $o_W = o_E = h_N \cup h_S$.

Problem Statement. The goal of this paper is to compute the game-theoretical value of the game Klaverjas KLAVERJASOPEN(h, ϕ, τ) with hands h, trump suit ϕ and starting player τ, where the players have full information. Team NS aims to maximize the the score of team EW subtracted from the score of team NS. Team EW aims to minimize this value. In case of a positive value, team NS has an optimal strategy to win the game, whereas in case of a negative value, team EW has an optimal strategy to win the game.

KLAVERJASOPEN(h, ϕ, τ) returns tuple $K = (K_{NS,t}, K_{NS,m}, K_{EW,t}, K_{EW,m})$ of four values, respectively the trick points obtained by team NS, the meld points obtained by team NS, the trick points obtained by team EW and the meld points obtained by team EW. As such, the total score of team NS can be defined as $K_{NS} = K_{NS,t} + K_{NS,m}$, and similarly the score of team EW can be defined as $K_{EW} = K_{EW,t} + K_{EW,m}$. Note that there are sometimes multiple sequences of moves may lead to the same outcome. Including the definition of *nat*, the result of the game is determined by OUTCOME(KLAVERJASOPEN(h, ϕ, τ)) =

$$\text{OUTCOME}(K) = \begin{cases} K_{NS} - K_{EW} & \text{if } K_{NS} > K_{EW}, \\ -(162 + K_{EW,m}) & \text{otherwise.} \end{cases}$$

This value is to be maximized by team NS and minimized by team EW. Note that team NS needs to obtain more points that team EW cf. the definition in Sect. 1, otherwise all 162 points are awarded to team EW (i.e., team NW is *nat*). In this case, the optimal outcome does never include any meld points from team NS, as the team obtaining meld points can choose to not declare these. Also note that the definition of *pit* from Sect. 1 is implicitly captured in this formalization through the meld points.

4   Exact Approach

After briefly discussing the combinatorics behind solving the game of Klaverjas in Sect. 4.1, after which the main approach is outlined in Sect. 4.2 and explored further in Sect. 4.3.

4.1   Combinatorics

In this section we restrict ourselves, without losing generality, to configurations of the game of Klaverjas with a fixed trump ($\diamondsuit$) and starting player (N). The number of different configurations is given as the number of ways of dealing 32 cards over 4 hands of 8 cards: $\binom{32}{8}\binom{24}{8}\binom{16}{8}$. Note that given a fixed trump suit some of the above configurations will be equivalent as the order of the non-trump suits is of no consequence. We omit the removal of these symmetrical configurations for the sake of simplicity.

We use a *Combinatorial Number System* [1,11] of degree k to define a bijection between any number N to the k-th combination (in lexicographic order) of $\binom{n}{k}$:

$$N = \binom{c_k}{k} + \cdots + \binom{c_2}{2} + \binom{c_1}{1}.$$

The process of mapping the number N to its corresponding combination is commonly referred to as *unranking*, while the inverse operation is called *ranking*.

It is trivial to combine combinatorial number systems (of the same degree). This allows us to easily enumerate the total number of configurations. Moreover, the total number of configurations is less than 2^{64} making the implementation trivial.

4.2 Solving Approach

Calculating the game-theoretical value of a given configuration with perfect information, a fixed trump and starting player can be done with various existing techniques, in particular minimax search with $\alpha\beta$-pruning [12]. In practical approaches, this search technique is often equipped with an intermediate evaluation function that allows for partial exploration of the search tree as well as various heuristics aimed towards reducing the statespace. In contrast to most practical implementations, we are interested in the game-theoretical value of a configuration, the search procedure needs to traverse the complete statespace (unless it can theoretically determine for a given branch that it will never be relevant, cf. $\alpha\beta$-pruning). In our implementation we use the classical minimax search with $\alpha\beta$-pruning, but without any additional heuristics. This approach is rarely practical as the statespace for most games is too large.

An upper bound on the statespace for a given configuration is $8!^4$, where in every trick each player is able to select any of his cards in hand. Although rare, this situation can occur in practice, e.g., when every player is dealt cards from only one suit.

The distribution of the suits over the hands is the main contributing factor to the number of legal moves during a particular game (note that in case of the trump suit also the face value of the card can be of influence). Therefore, the size of the search space highly depends on this distribution.

4.3 Equivalence Classes

In order to show the practicality of the solving approach presented in Sect. 4.2, we partition the total number of configurations into configurations with the same distribution of the number of cards from the same suit in each of the hands:

$$
\begin{pmatrix} 8\,0\,0\,0 \\ 0\,8\,0\,0 \\ 0\,0\,8\,0 \\ 0\,0\,0\,8 \end{pmatrix}, \ldots, \begin{pmatrix} 2\,2\,2\,2 \\ 2\,2\,2\,2 \\ 2\,2\,2\,2 \\ 2\,2\,2\,2 \end{pmatrix}, \ldots, \begin{pmatrix} 0\,0\,0\,8 \\ 0\,0\,8\,0 \\ 0\,8\,0\,0 \\ 8\,0\,0\,0 \end{pmatrix},
$$

where the hands are represented in rows and the columns represent the number of cards of each suit (both the order of players as well as the order of suits is unimportant). There are 981,541 of such equivalence classes[1], each containing a highly variable number of equivalent configurations ranging from 1 to 40,327,580,160,000. Note that equivalence does not imply a similar play out of the game nor a similar outcome. It merely fixes the distribution of cards of certain suits over the hands with the intent of obtaining a similar average branching factor during the $\alpha\beta$-search.

This partitioning focuses on exploring the effect of the various statespace sizes on the performance of our exact solving approach as well as yielding an interesting dataset for the machine learning approach described in Sect. 5.

5   Machine Learning Approach

In this section, we elaborate on our machine learning approach to predict the outcome of a particular configuration.

The classification problem is as follows. Given full information about the hands of all players, predict whether the starting team will win or not, i.e., whether for a given deal h, OUTCOME(KLAVERJASOPEN(h, ϕ, τ)) > 0. Like in Sect. 4, we fix $\phi = \Diamond$ and $\tau = N$.

In order to apply supervised machine learning techniques to this problem, we need to have access to a dataset of generated games, and their outcome. For this we use all 981,541 games that were analyzed in Sect. 6.1. Generally, machine learning models are induced based on a dataset $\mathcal{D} = \{(\mathbf{x}_i, y_i) \mid i = 1, \ldots, n\}$ to map an input $\mathbf{x}$ to output $f(\mathbf{x})$, which closely represents y. Here, n represents the number of games in the dataset, $\mathbf{x}_i$ is a numerical representation of one such game and y_i is the outcome of that game. As such, y_i represents whether team NS will obtain more points than team EW, i.e, whether OUTCOME(KLAVERJASOPEN(h, ϕ, τ)) > 0.

The main challenge is representing a game g as feature vector $\mathcal{F}(g)$. This has been done for other games, see for example Dou Shou Qi [20]. We note that more complex algorithms, e.g., convolutional neural networks, can implicitly learn this mapping. However, it has been noted in [8] that card games do not have a clear topological structure to exploit. As such, defining convolutions on card games is a research question in its own right and beyond the scope of this work. Table 2 shows an overview of the handcrafted features that we defined. The first column defines a name for each group of features, the column 'size' denotes how many

---

[1] See also: N.J.A. Sloane. The On-Line Encyclopedia of Integer Sequences, https:// oeis.org. Sequence A001496.

of those features can be generated. The column 'parameters' defines how this number of features can be generated. The last column defines how each feature can be generated.

Table 2. Features constructed for the machine learning approach.

Name	Size	Parameters	Definition
card ownership	32	$\forall s \in S, \forall f \in F$	$p : \exists c : c \in h_p \wedge suit(c) = s \wedge face(c) = f$
suit counts	16	$\forall p \in P, \forall s \in S$	$|h_{p,s}|$
rank counts	32	$\forall p \in P, \forall r \in R$	$|\{c \in H_p : rank(c) = r\}|$
points	4	$\forall p \in P$	$\sum_{c \in h_p} points(c)$
stdev per player	4	$\forall p \in P$	$stdev(\forall s \in S : |h_{p,s}|)$
stdev per suit	4	$\forall s \in S$	$stdev(\forall p \in P : |h_{p,s}|)$
stdev (game)	1		$stdev(\forall p \in P, \forall s \in S : |h_{p,s}|)$
top cards	16	$\forall p \in P, \forall s \in S$	$|\{c \in H_{p,s} : rank(c') < rank(c) \vee suit(c') \neq suit(c)\}|$ with $c' \in o_p$

Card ownership is a perfect mapping from a configuration containing all cards to a tabular representation. It describes for each card to which player it belongs. Suit counts, rank counts and points represent some basic countable qualities of the hand that a given player has. The standard deviation (stdev) gives a measure of how equal the suits are spread (per player, per suit and for the whole game). Note that the maximum obtainable standard deviation (per player, per suit and per game) is 3.46. If the game has a standard deviation of such value, this means that all players have all cards from a given suit (and the player with the cards from the trump suit will win the game). If the game has a standard deviation of 0, this means that all players have exactly two cards of each suit. The top cards denote how many cards of a given suit a player has that can not be beaten by the other team (except using trump cards).

Additionally, we can construct several convenience features oriented on teams, that can be exploited by the classifier to assess the card quality of the teams at once. For suit count (per suit), rank count (per rank), points and top cards (per suit), the appropriate team feature is the sum of both players for that feature. For example, the feature 'suit count of $\Diamond$ for team NS', is the sum of 'suit count of $\Diamond$ for player N' and 'suit count of $\Diamond$ for player S'.

6   Experiments

In this section we present results of using the exact approach and the machine learning approach, as well as a comparison between the approaches.

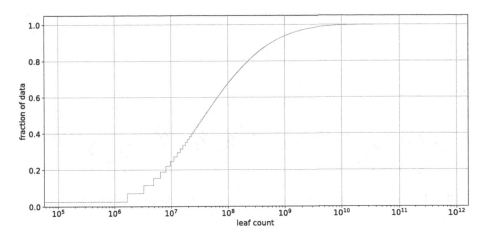

Fig. 1. Cumulative Distribution Function (CDF) of the leaf count of the exact algorithm for each of the 981,541 instances (see Sect. 4.3) of Klaverjas.

6.1 Exact Approach Results

As discussed in Sect. 4, the number of configurations is $\binom{32}{8}\binom{24}{8}\binom{16}{8}$, which is approximately $9.956 \cdot 10^{16}$. Therefore it is nontrivial to solve a representative sample of all configurations. Instead, we sample according to the equivalence classes as defined in Sect. 4.3. From each equivalence class we randomly select one configuration yielding a set of 981,541 configurations. For all of these configurations the game-theoretical score is calculated using minimax with $\alpha\beta$-pruning. No transposition tables were used. Note that when the team of the lead player can no longer obtain more than half of the points, all points will be assigned to the other team. Note that the leaf count within an equivalent class can differ significantly due to (i) the rule that a player needs to play a higher trump card if possible and (ii) the dynamics of $\alpha\beta$-pruning combined with how the cards are delt.

Figure 1 shows a CDF of the leaf count of the $\alpha\beta$-algorithm. This leaf count is directly proportional to the running time. Most configurations can be calculated in on average 1.5 CPU minutes, having between 10^6 and 10^9 leafs. A few instances with around 10^{11} leafs took around 45 minutes. However, one configuration required up to 4 CPU days, visiting $1.5 \cdot 10^{12}$ leafs.

6.2 Machine Learning Results

In this section we evaluate the performance of the machine learning techniques on the task to predict for a given configuration whether the team of the lead player can obtain more points than the other team. Our main interests are to evaluate the handcrafted features (as proposed in Sect. 5) and to compare machine learning approaches with exact search techniques (see Sect. 6.3). We use standard machine learning techniques, i.e., decision trees and random forests [3].

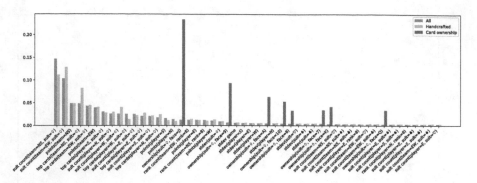

Fig. 2. Gini importance according to the random forest classifier. (Color figure online)

These have the advantage that they are interpretable and relatively insensitive to hyperparameter values. We use random forests with 64 trees. The other hyperparameters were set to their defaults, as defined in scikit-learn 0.20.0 [18].

For each configuration, we extract the following sets of features: card ownerships (first row in Table 2), the handcrafted features (all other rows in Table 2 and convenience team features) and all features (all rows in Table 2 and convenience team features). Clearly, the set of all features contains a richer source of information than the set of just the card ownerships. We evaluate the algorithm using 2-fold cross-validation. The problem is well-balanced (528,339 positive vs 453,202 negative observations). We record the predictive accuracy of the classifiers on all three feature sets. Predictive accuracy is the percentage of correctly classified observations.

The results are presented in Table 3, displaying the accuracy for different algorithms and feature sets. We note the following observations. As expected, the set of all features outperforms the set of just the card ownership features on both classifiers. Interestingly, the performance of the single decision tree on the handcrafted (and all) features almost equals that of random forest on just the card ownership features. Finally, the set of handcrafted features outperforms the set of all features for both classifiers. These observations lead us to belief that the handcrafted features on their own provide a more useful signal to learn from, in the context of tree-based models.

Table 3. Accuracy of machine learning algorithms on different feature sets.

Feature subset	Decision tree	Random forest
Card ownership	82.44	88.16
Handcrafted	88.02	91.98
All	87.96	91.79

In order to study the behaviour of the classifier a bit better, we analyze the Gini importance as defined in [4]. Gini importance is defined as the total decrease in node impurity (averaged by all trees in the ensemble). Intuitively, a high Gini importance resembles that the feature was important for classification. Figure 2 shows a bar plot of the 50 most important features (sorted according to the feature set 'All'). We show Gini importance for the three feature sets. Note that each feature is applicable to either the 'card ownership' set (blue bars) or the 'handcrafted set' (green bars). As a limitation of this feature importance analysis, we note that the notion of feature importance is rather subjective, and that there is no guarantee that this directly correlated with the predictive accuracy of a model.

The results seem to confirm several expected patterns. First, from the hand-crafted features that focus on a suit, the highest rated ones focus on the trump suit ($\Diamond$). Second, from the card ownership features, the highest rated ones are the ones that focus on the top two trump cards ($\Diamond$J and $\Diamond$9). Finally, from the rank count features, the highest rated ones focus on the highest rank (rank 8).

We note the following observations. First, when provided with all features, the handcrafted features provide the highest Gini importance. Second, the random forest makes proper use of the convenience team features (cf. top three features). Finally, suit counts, top cards and points seem to be strong features, often used in the top ranked features.

6.3 Comparison of Exact Approach and Machine Learning Approach

The two experiments above highlight how an exact algorithm based on $\alpha\beta$-search as well as a machine learning approach are both independently able to assess whether a certain hand of Klaverjas is winning. A comparison of which approach is "better" in terms of determining whether a hand is winning may at first glance seem uninteresting, as the exact algorithm always returns the correct answer. However, the number of leafs in the search tree that is traversed by the exact algorithm can be seen as an indicator of how difficult it is to exactly determine whether a hand is winning. A distribution of these leaf counts was presented in Sect. 6.1. Here, we compare this leaf count between four result sets from the machine learning model, namely the correctly classified win and loss instances, and the incorrectly classified win and loss instances.

Figure 3 presents results of this comparison. The figure shows on the vertical axis the number of leafs that were traversed by the exact algorithm, for all instances in each of the four result classes described above. Horizontal Gaussian noise was added to highlight the density of each result set at different leaf count values. Note the logarithmic vertical axis.

From the figure, we see that indeed, as discussed in Sect. 6.2, the majority of instances is correctly classified. More importantly, we observe two interesting patterns for the win instances (depicted in red and blue). First, it appears that the win instances require exploration of a larger search space than lost instances of the game. We believe that this is due to the fact that such instances require

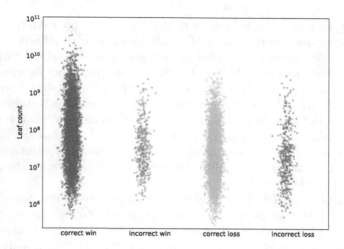

Fig. 3. Leaf count (according to exact algorithm) of correctly and incorrectly classified (according to machine learning approach) win and loss instances (1% sample). (Color figure online)

the algorithm to explore the maximum score up until the last trick of the game, whereas for lost games, the search tree can be pruned much earlier in the game when no more cards of substantial value are left. Second, we observe how for the correctly classified win instances (depicted in red), the number of leafs is significantly higher (note the logarithmic vertical axis). In fact, for incorrectly classified win instances (depicted in green) only substantially lower leaf counts are observed. It turns out that the machine learning algorithm is able to correctly classify the instances that are difficult to solve exactly (requiring up to four days of computation time, see Sect. 6.1). One possible explanation for this is the different objective of both approaches. Given a deal, the exact approach aims to find the set of moves that leads to the best possible score, whereas the machine learning approach only aims to classify whether a deal is winning or not. In future work we aim to study this in more detail, for example by comparing the exact approach against a supervised regression model.

7   Conclusion

In this paper, we have presented both an exact algorithm as well as a machine learning approach to solving the game of Klaverjas. In particular, we addressed the task of assessing whether a given distribution of cards over the hands of teams of players, is winning. It turns out that an exact algorithm based on $\alpha\beta$-search is able to determine this on average in a matter of minutes. In addition, the proposed machine learning approach employing simple features based on card ownership is able to predict whether a hand is winning with 88% accuracy. Adding more complex aggregated features and statistics related to meld points increases this accuracy to almost 92%. Interestingly, many of the cases where the machine learning algorithm consistently performs well, are in fact instances where the computation time (as a result of the number of leafs in the search tree) of the exact algorithm is longest (up to several days), evaluating over $1.5 \cdot 10^{12}$ play outs. This suggests that games that are difficult to assess for an algorithm, are in fact easy for humans, who typically use features similar to the machine learning approach in their decision making.

The findings presented in this paper highlight how in the future, a real-time artificial agent playing the game of Klaverjas may benefit from combining an exact approach with a machine learning approach, depending on what type of hand it is evaluating. A key question to then address is how we can a priori determine which component of such a hybrid algorithm we should use. The explored partitioning of the set of equivalence classes presented in this paper may provide a first basis of determining this.

In future work, we want to investigate if we can determine whether a hand is winning based on only one player's cards, rather than based on perfect information on all of the team's cards. A starting point may be the PIMC approach discussed in [15], where to solve a game with imperfect information, typically many games with perfect information are played out. In addition, we want to investigate the use of transposition tables in order to reduce the search space even further.

Acknowledgements. The second author was supported by funding from the European Research Council (ERC) under the EU Horizon 2020 research and innovation programme (grant agreement 638946). We thank F.F. Bodrij and A.M. Stawska for assistance with qualitative real-world validation of a relevant feature subset.

References

1. Beckenbach, E.F.: Applied Combinatorial Mathematics. Krieger Publishing Co., Inc., Melbourne (1981)
2. Bonnet, É., Jamain, F., Saffidine, A.: On the complexity of trick-taking card games. In: IJCAI, pp. 482–488 (2013)
3. Breiman, L.: Random forests. Mach. Learn. **45**(1), 5–32 (2001)
4. Breiman, L., Friedman, J., Stone, C.J., Olshen, R.: Classification and Regression Trees. Chapman and Hall/CRC, Wadsworth (1984)
5. Buro, M., Long, J.R., Furtak, T., Sturtevant, N.R.: Improving state evaluation, inference, and search in trick-based card games. In: IJCAI, pp. 1407–1413 (2009)
6. Frank, I., Basin, D.: Search in games with incomplete information: a case study using bridge card play. Artif. Intell. **100**(1–2), 87–123 (1998)
7. Ginsberg, M.L.: GIB: imperfect information in a computationally challenging game. J. Artif. Intell. Res. **14**, 303–358 (2001)
8. Hearn, R.A.: Games, puzzles, and computation. Ph.D. thesis, Massachusetts Institute of Technology (2006)
9. van den Herik, H.J., Uiterwijk, J.W., van Rijswijck, J.: Games solved: now and in the future. Artif. Intell. **134**(1–2), 277–311 (2002)
10. Hoogeboom, H.J., Kosters, W.A., van Rijn, J.N., Vis, J.K.: Acyclic constraint logic and games. ICGA J. **37**(1), 3–16 (2014)

11. Knuth, D.E.: The Art of Computer Programming, Volume 4, Fascicle 3: Generating All Combinations and Partitions. Addison-Wesley Professional, Boston (2005)
12. Knuth, D.E., Moore, R.W.: An analysis of alpha-beta pruning. Artif. Intell. **6**(4), 293–326 (1975)
13. Kupferschmid, S., Helmert, M.: A skat player based on Monte-Carlo simulation. In: van den Herik, H.J., Ciancarini, P., Donkers, H.H.L.M.J. (eds.) CG 2006. LNCS, vol. 4630, pp. 135–147. Springer, Heidelberg (2007). https://doi.org/10.1007/978-3-540-75538-8_12
14. Long, J.R., Buro, M.: Real-Time opponent modeling in trick-taking card games. In: IJCAI, vol. 22, pp. 617–622 (2011)
15. Long, J.R., Sturtevant, N.R., Buro, M., Furtak, T.: Understanding the success of perfect information Monte Carlo sampling in game tree search. In: AAAI (2010)
16. Parlett, D.: The Penguin Book of Card Games. Penguin, London (2008)
17. Pearl, J.: The solution for the branching factor of the alpha-beta pruning algorithm and its optimality. Commun. ACM **25**(8), 559–564 (1982)
18. Pedregosa, F., et al.: Scikit-learn: machine learning in Python. J. Mach. Learn. Res. **12**, 2825–2830 (2011)
19. van Rijn, J.N., Takes, F.W., Vis, J.K.: The complexity of Rummikub problems. In: Proceedings of the 27th Benelux Conference on Artificial Intelligence (2015)
20. van Rijn, J.N., Vis, J.K.: Endgame analysis of Dou Shou Qi. ICGA J. **37**(2), 120–124 (2014)
21. Silver, D., et al.: Mastering the game of Go with deep neural networks and tree search. Nature **529**(7587), 484–489 (2016)
22. Silver, D., et al.: Mastering the game of Go without human knowledge. Nature **550**(7676), 354–359 (2017)
23. Wästlund, J.: A solution of two-person single-suit whist. Electron. J. Comb. **12**(1) (2005). Paper #R43
24. Wästlund, J.: Two-person symmetric whist. Electron. J. Comb. **12**(1) (2005). Paper #R44

Style Transfer of Abstract Drum Patterns Using a Light-Weight Hierarchical Autoencoder

Mark Voschezang[(✉)][iD]

VU University, 1081 HV Amsterdam, The Netherlands
mark.voschezang@icloud.com

Abstract. Many improvements have been made in the field of generative modelling. State-of-the-art unsupervised models have been able to
transfer the style of existing media with photo-realistic quality. However,
these improvements have been largely limited to graphical data. Music
has been proven to be more difficult to model. Magenta's MusicVAE
can quite successfully generate abstract rhythms and melodies. However, MusicVAE is a large model that requires vast amounts of computing
power before it starts to make realistic predictions. Moreover, its input is
heavily *quantized* which makes it impossible to model musical variations
such as swing. This paper proposes a lightweight but high-resolution
variational recurrent autoencoder that can be used to transfer the style
of input samples while maintaining characteristics of the original sample.
This model can be trained in a few hours on small datasets and allows
researchers and musicians to experiment with musical style transfer. In
addition, a novel technique based on *normalized compression distance*
is used to evaluate the model by measuring the similarity of generated
samples to target classes.

Keywords: Variational autoencoder · MIDI drum patterns ·
Generative modelling · Normalized compression distance

1 Introduction

In the past years many improvements have been made in the field of generative
modelling. Examples of models that can produce realistic looking images are
CycleGAN, MUNIT and the model by Karras et al. (2017), which rely on many
convolutional layers [12,14,17]. Unlike convolutional layers, recurrent layers are
designed to make predictions based on previous input values and thus should be
better suited for temporal data such as music. Although some successful music-
generating models have been created, training networks that incorporate these
layers requires enormous amounts of computing power [24].

Music can be encoded in many different digital formats. Common high-
quality formats are FLAC and WAV, which store audio in a bitstream. Alternatively, music can be represented in an abstract notation such as MIDI [23].

© Springer Nature Switzerland AG 2019
M. Atzmueller and W. Duivesteijn (Eds.): BNAIC 2018, CCIS 1021, pp. 121–137, 2019.
https://doi.org/10.1007/978-3-030-31978-6_10

Comparable to sheet music, MIDI is an instrument-invariant format that contains only the most essential musical information that defines a piece. A MIDI file consists of a stream of note-messages that can be played back by a synthesizer or sampler. As a single MIDI file can be played back by different instruments the resulting timbre of the audio can differ drastically.

Magenta's MusicVAE is a variational autoencoder that can reconstruct MIDI-based drum patterns and melodies [27]. It can apply latent-space transformations to encoded input samples to transform them to sound similar to a target sample. Decoding an interpolation of the latent space produces a musically related sequence of samples. A study showed that music generated by MusicVAE sounded comparable to music that was created by humans. However, the model has a number of limitations. (1) It contains has many trainable parameters and requires a large dataset to be trained on. Training the model requires a large amount of computing power. (2) During encoding of MIDI files to vector sequences that fit the input of the model, a quantization with a resolution of 16th notes is applied. This means that information about *swing, groove* and *feel* is lost. Especially these aspects are what makes drum patterns sound human and are what distinguishes different playing styles [6,8,32].

This paper proposes a model that is based on MusicVAE. However, it is much faster to train due to a reduced amount of parameters and the usage of both convolutional and recurrent layers. At the same time, it uses a higher quality of MIDI conversion to allow for subtle differences in timing.

The next section introduces a number of general concepts and techniques that are vital to this paper. This is followed by a description of our specific model and an experiment to test the musicality of the transformations that can be applied with this model. Finally, a thorough assessment of the assumptions of this model is given along with recommendations for further research.

2   Background

2.1   MIDI

A MIDI file consists of a stream of messages. Each message contains the pitch, velocity and duration of a note that occurs at a certain instance. Additionally, MIDI files can contain control- and meta-messages, but these are discarded in this paper. Temporal information is quantized to a musical grid, with a resolution that varies per file.

This means that a perfect timing can be maintained when changing the tempo of a song, but notes that played off-beat may be shifted to fit the grid. In drum patterns this happens when a certain level of swing is introduced, or when for example the snare drum is played deliberately early. In some musical styles drums are deliberately played with a sloppy (i.e., imperfect) timing [8,9]. This is done to achieve a certain (sometimes human-like) feel or to make the groove more dance-able.

Velocity-information of individual notes is quantized even heavier and can contain not more than 128 different values. With the exception of dedicated (drum-)synthesizers, samplers usually implement changes in velocity by increasing the volume at which samples are played. The resulting variation in dynamics is far less intensive than the dynamics that are caused by striking a physical instrument with higher velocity, which causes a change in *timbre* (i.e., audible frequencies) and decay of the sound. This lack of variation can make synthesized MIDI files sound artificial. Despite these limitations, MIDI is a widely used format. Carefully programmed patterns and melodies that are synthesized can sound as musical as audio that is produced by acoustic instruments.

There are two aspects of MIDI that make it especially suitable for generative modelling. Firstly, it requires less memory than waveform-based formats, such as WAV. Secondly, the abstract design of MIDI itself is a first step into finding generalizations of training data.

2.2   Variational Autoencoders

The model that is used is a type of Variational Autoencoder (VAE) [1,16]. An autoencoder consists of an encoder and a decoder [28]. An input x is encoded into a lower dimensional latent vector z, which is then decoded back to a reconstruction of the input x'. This allows an autoencoder to be trained on unlabelled data. This is especially useful in a domain such as music where classes (genres) are subjective; music is perceived differently between people [18]. The reduction in dimensionality of z forces the model to learn a generalized form of the training data.

A VAE is an autoencoder in which the encoding part predicts not the latent vector itself but instead the (conditional) multivariate probability distribution $P(z|x)$. The latent space is assumed to be a standard normal distribution ($P(z) \sim \mathcal{N}(0,1)$) [15,26]. The decoder models the conditional distribution $P(x'|z)$. This forces decoder to be more invariant to change. In practice, this technique stimulates the decoder to always predict a musical output (i.e., as musical as the training data) even if the decoder's input is random.

The encoder should not just encode samples of different genres into different latent vectors, but changes in input samples should correspond to proportional changes in the encoded representation of that sample. As a result, decoding a sequence of nearby latent vectors will result in a set of output samples that are similar. Decoding a movement in the latent space from vector z_a to vector z_b should result in a sequence of samples that will be perceived as a smooth transition from the original sample a to sample b.

2.3   Recurrent and Hierarchical Layers

Music consists of different patterns, that recur at different temporal positions in a piece. In traditional models this would mean that a feature detector (i.e., a dedicated set of neurons that are activated by some pattern in the input data) has to be repeated for every position where the pattern could occur. To prevent an unnecessary increase of trainable parameters two types of layers are used: recurrent layers and shared layers.

Recurrent layers allow a neural network to be able to make predictions based on previous input values [4]. Although simple recurrent neural networks often have trouble in modelling long term dependencies, *long short-term memory* (LSTM) based networks have proven to be able to learn both short and long-term dependencies [2,11]. This makes them suitable to model music.

Recurrent layers are very powerful but are generally more difficult to train than convolutional layers [9]. A simple alternative is to use shared layers [33]. Multiple copies of a single dense layer that share weights and biases are applied to different temporal slices on an input. This allows a shared layer to be applied to long input sequences, without requiring an increase in trainable parameters. This allows a model to learn to "recognize" features at different temporal locations, without requiring these features to be present at multiple temporal locations. In this sense, the usage of shared layers increases the level of generalization of a model. This is useful if a model has to be trained on a limited amount of data.

3   Model

This section covers the structure of the model. The model consists of an encoder that encodes matrix-representations of MIDI files to latent vectors and a decoder that applies random transformations to latent vectors and then generates reconstructions of the original input matrices. Figure 1 shows an abstract overview of the model.

3.1   Encoder

The encoder starts with a 1-dimensional convolutional layer, with the purpose of making the model invariant to small variations in the input data [7]. To reduce the loss of spatial information, 64 output filters are used in the convolutional layer. This layer is followed by a bidirectional LSTM layer with $2 \cdot 128$ nodes and a many-to-one encoding [10]. The output of the encoder are two parallel dense layers with 20 nodes that act as z_μ and $z_{log\sigma}$. These layers do have trainable weights and biased but lack activation functions.

3.2   Latent Space

The encoder outputs 2 latent vectors that are called z_μ and $z_{log\sigma}$. These are used to in the reparameterization of the *evidence lower bound* to obtain a random latent vector z [15,16,26]. This value is computed by (point-wise) multiplying a random vector ϵ (sampled from a multivariate normal distribution) with $z_{log\sigma}$ and adding the result to z_μ.

$$z = z_\mu + \epsilon \circ e^{z_{log\sigma}} \tag{1}$$

3.3 Decoder

The hierarchical structure of the generating part of the network is based on MusicVAE [27]. Instead of decoding the latent vectors directly as done in traditional autoencoders, they are fed into a *conductor-network*. This network outputs a sequence of embeddings, that are independently fed into a *embedding-decoder*. The output sequence of the embedding-decoder is reshaped to the shape of the original input. It is not proven that the specific structure of this model improves the performance of the model. Instead, it is used as a restriction that forces the model to learn a hierarchical representation of the training data. The first layer after re-parameterization is a dense layer with 256 nodes and a leaky-ReLU activation layer [20].

$$\text{leakyReLU}(x) = \begin{cases} 0.3x, & \text{if } x < 0 \\ x, & \text{otherwise} \end{cases} \tag{2}$$

This layer is followed by three bypassed dense layers with 256 nodes and elu activation [25]. These activation functions are added to speed up the training of the model.

$$\text{elu}(x) = \begin{cases} 0.1(e^x - 1), & \text{if } x < 0 \\ x, & \text{otherwise} \end{cases} \tag{3}$$

A summation layer merges the previous layer with the first layer of the decoder. The purpose of this layer is to initially bypass the three elu layers and start using them later when the model requires more complexity to keep improving [30]. Batch-normalization is used to keep the model more stable and decrease the training time further [13]. These layers are reshaped into a sequence of 10 vectors and are fed into a bidirectional LSTM layer with 2×128 layers [10,11]. This recurrent layer outputs a sequence of embeddings that are fed in the embedding-decoder.

The embedding-decoder starts with a dense layer with 250 nodes and a relu activation. The output layer of the embedding-decoder is a dense layer with sigmoid activation to keep the output values in the correct range. The length of this layer is equal to the product of the length of the embedding (i.e., the sample length divided by the amount of embeddings) and the amount of notes. Finally, all the embeddings are combined and reshaped to the original input shape.

3.4 Loss Function

The parameters of the model are optimized using the *Adam* optimizer [21] and a summation of different loss functions. The main component of the loss functions is the binary cross-entropy (H_C) of x and x'. Minimizing this part of the loss function causes the model to learn the identity function. The second component is the Kullbach-Leibler divergence (D_{KL}) of the predicted latent vectors and

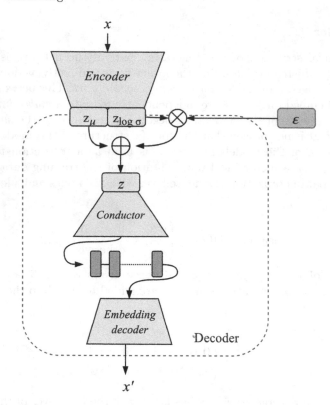

Fig. 1. The structure of the Variational Autoencoder. The input matrix x is fed into the *encoder* and encoded into the vectors z_μ and $z_{log\sigma}$ which represents the mean and variance of the (multivariate) conditional distribution $P(z|x)$. The latent vector z is computed by (point-wise) multiplying the variance $z_{log\sigma}$ with a random vector ϵ and adding the mean z_μ. The *conductor-network* uses z as input and produces a sequence of *embeddings*. These are decoded to an output matrix x'. The network is trained to generate an x' that is similar to an input x.

the standard normal distribution [15,31]. This forces the encoder's output to be normally distributed. Lastly the mean square error (MSE) between x and x' is added to allow the model to minimize non-binary differences in output, which corresponds to differences in the velocity of MIDI notes.

$$\text{loss}(x) = \alpha H_C(x, x') - \beta D_{KL}\big(P(z|x) \parallel P(z)\big) + \gamma \, \text{MSE}(x, x') \qquad (4)$$

4   Method

This section describes the process used to obtain training data and an experiment that tests the model's ability to generate musically meaningful transformations.

4.1   Training Dataset

The training dataset consists of 497 MIDI files from different internet sources and contains various musical styles, including pop, rock and jazz. There were a total number of 71 classes and 7 samples per class. Many classes may have been similar in a musical sense, but there was no convention between classes from different sources. Thus, it is highly likely that these classes are not mutually exclusive. Note that a class does not necessarily resemble a single musical genre.

This paper focuses on short drum loops. Samples with the word "fill" in the filename are omitted from the training set because drum fills may sound unmusical when played out of context. It is assumed that all samples have a tempo of 120 bpm and start on the first beat (instead of starting with an upbeat, which is not unusual), but this is not checked explicitly. As a result, the training set consists of *4-bar* drum samples, that each have length of 2.0 s at 120 bpm.

The model will be trained for 1000 epochs on an *Intel i5* CPU using the Adam optimizer with a fixed learning rate of 0.001 and a batch size of 128.

4.2   Encoding of MIDI Files

In MIDI drum patterns, the pitch-value of note-messages corresponds to the type of instrument instead of to the actual pitch. Not all MIDI files use the same pitch-instrument standard. For this reason, a lookup table is used to map different pitches to 9 distinct instruments. These instruments are: bass drum, snare drum, closed hi-hat, open hi-hat, low tom, middle tom, high tom, crash cymbal and ride cymbal. All notes that do not correspond to a known pitch are merged into an additional channel. All MIDI files are encoded into sequences of 160 note-vectors. A note-vector contains the velocities at which the different instruments are hit during the corresponding interval. An interval has a duration of 0.025 s. Velocities that are lower than 10% of the maximal velocity are interpreted as *rests* (silence) and are not converted.

Unlike many melodic instruments, most acoustic drums are played without an intended duration of notes. This means that they are triggered once and then left to ring out. For this reason, *note-off* messages are discarded during encoding of messages. There are a few exceptions such as the (open) hi-hat and crash cymbals. These instruments are occasionally muted manually by drummers. In such cases there is a small loss of musical information.

4.3   Evaluation

The extent to which the model can transfer the style of drum patterns is tested with the metric *normalized compression distance* (NCD) [19]. The similarity of transformed samples is compared to both the target and the original class. Note that transformations of style are different than transformations of a sample itself. The latter can be achieved perfectly, given that a model is trained on both the input and target sample. This paper focuses on the transfer of the style of samples while maintaining characteristic aspects of the original class.

Application of latent vectors. Application of latent vector transformations is defined as the point-wise addition of two vectors. Transformations can be applied with variable intensity by scaling the transformation vector α.

$$v_a(v_{\text{base}}) = v_{\text{base}} + \alpha v_a \tag{5}$$

Transformations. For every class in the training data, the average of all latent transformations from every sample in that class to the average latent location of all other classes is computed. This is considered the set of latent transformations that correspond with transforming the style of class $\mathcal{A}$ to the style of class $\mathcal{B}$ (abbreviated as $\mathcal{A} \rightarrow \mathcal{B}$).

$$\bar{b} = \frac{1}{|\mathcal{B}|} \sum_{i=1}^{|\mathcal{B}|} \mathcal{B}_i$$

$$\text{transformation}(\mathcal{A}, \mathcal{B}) = \frac{1}{|\mathcal{A}|} \sum_{i=1}^{|\mathcal{A}|} \bar{b} - \mathcal{A}_i \tag{6}$$

A *minimal transformation* is considered a transformation $\mathcal{A} \rightarrow \mathcal{B}$ only the most distinguishing latent dimension is modified. For every transformation the minimal transformation is approximated with an *ensemble* of decision tree classifiers [3, 26].

A number of transformations is randomly sampled from the set of transformations described in the previous section. Each transformation $\mathcal{A} \rightarrow \mathcal{B}$ is applied to all samples of class A. The NCD between the transformed set and both the original set and the target set is computed. This is repeated with several transformation intensities. The experiment is performed for full and for minimal transformations.

It is expected that changes in the latent space map to proportional changes in the NCD between the original samples and the samples in the target class. Thus, for every transformation $\mathcal{A} \rightarrow \mathcal{B}$ there will be a negative correlation between the transformation intensity and the NCD between the transformed samples $\mathcal{A}'$ and the target samples in $\mathcal{B}$. As the minimal transformations are less invasive, it is expected that the correlation between minimally transformed samples and samples of the corresponding original classes is lower than the correlation of the fully transformed samples and the original classes.

5   Results

Figure 2 shows the training loss during training. The figure suggests that the model was still learning when the training was stopped. It was assumed the model was overfitting but this was not checked explicitly.

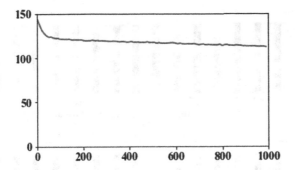

Fig. 2. The training loss for every epoch of training, computed with Eq. (4).

5.1 Qualitative Evaluation

Manual inspection of transformations showed that not all transformations had an audible effect (i.e., there were no notes added or removed). This was the case for both full and minimal transformations. It seems that the magnitude to which a latent transformations modifies the style of a latent vector, is depending on the original latent position of that vector.

Figure 3 shows an example of a "successful" transformation sequence. At each sequential step, a small number of notes of the previous pattern are changed. This shows that the model is capable of generating subtle transformations.

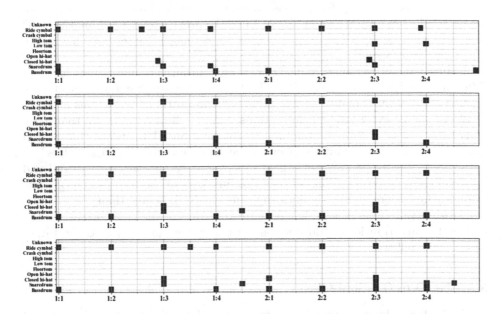

Fig. 3. A number of transformed patterns. The original class (label) is *"country - straight brushes"* and the target class is *"jazz - another you"*. The figure shows the first 2 beats of the patterns (there are 4 beats per bar). Each pattern was automatically denoised by removing notes that occured within 3 intervals after each note.

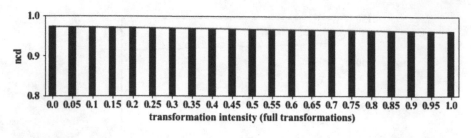

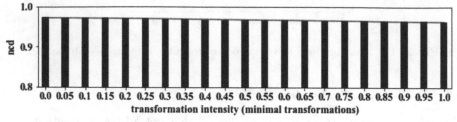

Fig. 4. The average NCD between transformed classes and target classes with increasing intensities for full transformations (above) and minimal transformations (below). The line is the average of the regression lines of each transformation. Note that the y-axis does not start at 0.

5.2 NCD to Target Class

For every class, 21 transformations to random other classes were computed, which resulted in a total sample size of 1491. For every transformation, the NCD was applied as described in Sect. 4.3, with 20 incremental intensities between 0 and 1 per transformation.

Figure 4 shows the average NCD between transformed and target classes of different transformation intensities. The figures suggests that the intensities of both types of transformations correspond with proportional reductions in NCD. 58.3% of the slopes of full transformations were significantly different than zero. This value was 38.8% for the minimal transformations. This means that not all transformations had a positive effect on the similarity with the target class.

A dependent t-test was performed independently for every pair of full and minimal transformations. This showed that in 43.9% of the transformations, the full transformations led to significantly different NCD scores than the minimal transformations. This means that in more than half of the transformations the minimal transformations produced samples that were as similar to the target class as the full transformations.

5.3 NCD to Original Class

Figure 5 shows the average NCD between transformed and original classes of different transformation intensities. The figures suggests that the intensities of both types of transformations correspond with proportional growth in NCD. For

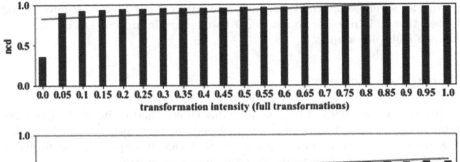

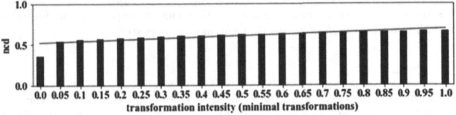

Fig. 5. The average NCD between transformed classes and original classes with increasing intensities for full transformations (above) and minimal transformations (below). The line is the average of the regression lines of each transformation.

the full transformations 70% of the slopes were significantly different than zero. This value was 89% for the minimal transformations. This means that not all transformations decreased the similarity with the original class gradually.

A dependent t-test was performed independently for every pair of full and minimal transformations. This showed that the full transformations led to significantly higher NCD scores that the minimal transformations in 82% of the transformations. This means that in 82% of the transformations the minimal transformations were less invasive.

6   Conclusion

The experiment showed that the proposed model can successfully generate latent transformations that modify the style of drum patterns towards a target style, with variable intensity. It also showed that the minimal transformations were better at maintaining similarity of the transformed samples to the original classes than the full transformations. Not all latent transformations affected the style of pattern with the same intensity. This suggests that the model did not learn to use all latent dimensions to the same extend. There could be several reasons for this. (1) The training of the model was stopped too early. (2) There was not enough variance in the training dataset with respect to the size of the model. This means that the model could reconstruct all training data without needing to generalize the data to a certain extend.

The techniques described in Sect. 4.3 can be used to automatically extract meaningful transformations for some classes, but this method is not guaranteed to work for all classes. Finally, the implementation of the NCD-based evaluation showed that it is possible to use NCD to measure improvements in similarity between multiple samples.

7   Further Research

During the development of this model many choices had to be made about the training data, the structure of the model and the transformations of samples. This section gives an overview of aspects that can be improved and probably will enhance the transformations that the model generates.

7.1   Dataset

The model was trained on a small dataset, especially compared to the size datasets that were used to train MusicVAE and MUNIT [7,12]. While the ability to extract multiple musical properties from a limited amount of data is desired, the variance of the used dataset may have been too small with respect to the size of the model and the amount of training time. Using a larger dataset without increasing the size of the model will force the model to learn more subtle differences between samples and this may result in overall better transformations (i.e., higher absolute correlations between transformation intensity and the NCD of transformed samples to both the target and original classes).

Another possible change is to not increase the size of the dataset, but to use a smaller number of classes. Training a model on less classes will reduce the generalizability of a model, which effectively means that the reconstruction quality of new, unseen data will be lower. As long as the model is used to transform samples from within its training dataset, this is not a problem. The advantage of training a model on fewer classes is that more subtle differences between classes will be modelled. It can be expected that this will improve the model's ability to apply subtle transformations.

It is also advised to take the musical compatibility of different samples into account. Samples in the training dataset were cropped to have a length of 2 bars and speed up or down to have a tempo of 120 bpm. Using samples that have comparable lengths and tempi will reduce the (possibly non-musical) side-effects of "normalizing" the training data.

7.2   Model Structure

Multiple versions of embedding-decoders were tried during the development of this model, including dense, convolutional and recurrent layers [34]. We found that using dense or convolutional layers reduced the training time, and that dense embedding-decoders produced a less "blurry" output than convolutional layers. However, the effect that these different embedding-decoders had on the quality of generated transformations was not tested.

It is possible that the model was complex enough to memorize a significant part of the training data, which reduced the level of generalization of the model. The most probable aspects of the model structure that would instigate this are the size of the latent space (i.e., the amount of latent dimensions) and the group of dense layers in the conductor decoder. The latent space is the part of the model that has the lowest dimensionality and thus determines the ultimate amount of compression of input data. The dense layers in the conductor decoder together account for a significant of the trainable parameters of the model and can store a lot of complexity.

It is desired to have a high-dimensional latent space, as the purpose of latent dimensions is to correspond to the many different musical properties of the training data. Therefore it is advised to reduce either the amount of nodes in the dense layers, or the amount of dense layers.

The hyperparameters of the model were not optimized. In MusicVAE, several loss function parameters were altered during training [27]. This technique can be used for this model as well. The loss function (as described in Sect. 3.4) could initially be biased by using a high value for α and a low value for β. Inverting the relation over time will stimulate the model to first learn to reconstruct the input and later learn a generalized form of the input. Increasing γ instead of β will allow the model to first learn reconstruction that is invariant to velocity, and later learn more subtle differences.

7.3 Noise

A major shortcoming of this model is that the prediction occasionally contains stuttering of notes and an unusual co-occurrence of notes. Such "mistakes" can sound glitchy and artificial. The stuttering of notes can simply be a side-effect of different patterns that are combined. The co-occurrence of notes may have a more complicated cause.

The co-occurrence of notes seemed to happen for instruments that sound similar, for example a floor tom, low tom and high tom. Because these instruments sound similar, they can often be interchanged (within a musical piece). Therefore, it is likely that they occur often in similar positions in patterns. It is expected that the model has not learned that these instruments will have high independent probabilities to occur in a certain pattern, but have a much lower probability to occur together in that pattern (i.e., $P(e_1|x) \gg P(e_1|e_2, x)$, for events $e_{1,2}$ in pattern x).

Although it should be possible for the main model to learn these dependencies, it could be better to build a separate model that is focused on removing this kind of noise. This separate model could be a *denoising* autoencoder [29]. Adversarial examples could be generated by adding random noise to samples or by randomly adding samples from different classes together. These techniques have successfully been used to build denoising autoencoders for images, and can be expected to work to some extend for MIDI matrices.

7.4   Transformations

The model can be used to an even fuller extend by defining more complex transformations. An alternative to selecting the single most distinguishing latent dimension, as done in minimal transformations, is to transform every latent dimension proportionally to the relative importance of that dimension for two classes. This method should have the advantages of both full and minimal transformations.

The style of samples can also be influenced directly, by using a model that is structured as a *conditional* variational autoencoder (CVAE) [5]. A CVAE extends a VAE by modelling auxiliary label information. This allows the training dataset to be extended with domain knowledge. Style transfer can be applied by decoding the combination of an encoded input sample and an altered label-vector. It will be interesting to how the transformations produced by the two approaches will differ.

7.5   Metrics

The implementation of NCD that was used to determine similarity of patterns to classes was based on the generic compression algorithm *zip* [19]. More adequate compression algorithms are SIA and COSIATEC [22]. These algorithms are designed to compress abstract music and can be expected to improve the extent to which NCD measures similarity in a musical sense.

A future use case for NCD is to use it during hyper parameter optimization of generative models. Commonly used loss functions (including the loss functions described in Sect. 3.4) perform well when measuring similarity of individual samples, but are not designed to recognize similarity to sets of samples. NCD can be used to measure the similarity of predicted output samples to target classes, even though the predicted samples are not equal to a specific output sample. This makes NCD suitable to be incorporated in a heuristic that focuses on learning parameters that are optimal for transferring the style of input data, as style transfer is a fuzzy process that not aims to produce a perfect reconstruction of training data, but rather to generate a partial reconstruction that contains specific characteristics of certain data.

Acknowledgments. The author wishes to thank Stefan Schlobach, Albert Meroño Peñuela and Peter Bloem for inspiration and useful discussions.

A   Appendix

Both the implementation of the model described in this paper and a number of synthesized examples of generated MIDI files can be found at https://github.com/voschezang/drum-style-transfer.

A.1   Parameters

Table 1 shows the values of the most important parameters.

Table 1. Parameters

Parameter	Value
Dimensionality of the latent space	10
Input shape (timesteps, notes)	(160, 10)
Maximal quantization	0.025 s
Training dataset	497 samples
Amount of classes	72
α (binary crossentropy loss intensity)	1
β (KL loss intensity)	0.75
γ (MSE loss intensity)	0.05
Batch size	128
Number of training epochs	1000

A.2   Structure of the Model

The encoder and decoders can be seen as a pipeline where a sequence of transformations is applied to an input. Table 2 shows a brief overview of each layer.

Table 2. Structure of the model

Encoder		
0	(input)	64 output filters, ReLU, kernel size − 2, strides = 1
1	conv1d	128 nodes, activation functions as explained in [17]
2	bidirectional LSTM	(both dense, no activation layers)
3	(z-mean, z-log-var)	
Decoder		
0	(input)	produces a list of embeddings
1	reparameterization	is mapped to each embedding
2	**Conductor network**	reshape list of embeddings
3	**Embedding decoder**	
4	output	
Conductor network		
0	(input)	
1	dense 1	*leaky ReLU* activation
2	dense 2–5	*elu* activation
3	sum(dense 1, dense 5)	(bypass for layers 2–5)
4	batch normalization	
5	repeat-vector	(10 times)
6	BiLSTM	(return sequences (embeddings))
Embedding decoder		
0	(input)	
1	dense	*ReLU* activation
2	dense	*sigmoid* activation

References

1. Baldi, P.: Autoencoders, unsupervised learning, and deep architectures. In: Proceedings of ICML Workshop on Unsupervised and Transfer Learning, pp. 37–49 (2012)
2. Bellec, G., Salaj, D., Subramoney, A., Legenstein, R., Maass, W.: Long short-term memory and learning-to-learn in networks of spiking neurons. arXiv preprint arXiv:1803.09574 (2018)
3. Breiman, L.: Random forests. Mach. Learn. **45**(1), 5–32 (2001)
4. Cho, K., et al.: Learning phrase representations using RNN encoder-decoder for statistical machine translation. arXiv preprint arXiv:1406.1078 (2014)
5. Creswell, A., Bharath, A.A., Sengupta, B.: Conditional autoencoders with adversarial information factorization. arXiv preprint arXiv:1711.05175 (2017)
6. D'Errico, M.A.: Behind the beat: technical and practical aspects of instrumental hip-hop composition. Ph.D. thesis, Tufts University (2011)
7. Dumoulin, V., Visin, F.: A guide to convolution arithmetic for deep learning. arXiv preprint arXiv:1603.07285 (2016)
8. Fujii, S., Hirashima, M., Kudo, K., Ohtsuki, T., Nakamura, Y., Oda, S.: Synchronization error of drum kit playing with a metronome at different tempi by professional drummers. Music Percept.: Interdiscip. J. **28**(5), 491–503 (2011)
9. Gers, F.A., Schraudolph, N.N., Schmidhuber, J.: Learning precise timing with LSTM recurrent networks. J. Mach. Learn. Res. **3**(Aug), 115–143 (2002)
10. Graves, A., Fernández, S., Schmidhuber, J.: Bidirectional LSTM networks for improved phoneme classification and recognition. In: Duch, W., Kacprzyk, J., Oja, E., Zadrożny, S. (eds.) ICANN 2005. LNCS, vol. 3697, pp. 799–804. Springer, Heidelberg (2005). https://doi.org/10.1007/11550907_126
11. Hochreiter, S., Schmidhuber, J.: Long short-term memory. Neural Comput. **9**(8), 1735–1780 (1997)
12. Huang, X., Liu, M.Y., Belongie, S., Kautz, J.: Multimodal unsupervised image-to-image translation. arXiv preprint arXiv:1804.04732 (2018)
13. Ioffe, S., Szegedy, C.: Batch normalization: accelerating deep network training by reducing internal covariate shift. arXiv preprint arXiv:1502.03167 (2015)
14. Karras, T., Aila, T., Laine, S., Lehtinen, J.: Progressive growing of gans for improved quality, stability, and variation. arXiv preprint arXiv:1710.10196 (2017)
15. Kingma, D.P., Salimans, T., Jozefowicz, R., Chen, X., Sutskever, I., Welling, M.: Improved variational inference with inverse autoregressive flow. In: Lee, D.D., Sugiyama, M., Luxburg, U.V., Guyon, I., Garnett, R. (eds.) Advances in Neural Information Processing Systems 29, pp. 4743–4751. Curran Associates, Inc. (2016). http://papers.nips.cc/paper/6581-improved-variational-inference-with-inverse-autoregressive-flow.pdf
16. Kingma, D.P., Welling, M.: Auto-encoding variational bayes. arXiv preprint arXiv:1312.6114 (2013)
17. Liao, J., Yao, Y., Yuan, L., Hua, G., Kang, S.B.: Visual attribute transfer through deep image analogy. arXiv preprint arXiv:1705.01088 (2017)
18. Lippens, S., Martens, J.P., De Mulder, T.: A comparison of human and automatic musical genre classification. In: 2004 Proceedings of IEEE International Conference on Acoustics, Speech, and Signal Processing (ICASSP 2004), vol. 4, pp. iv-233–iv-236. IEEE (2004)
19. Louboutin, C., Meredith, D.: Using general-purpose compression algorithms for music analysis. J. New Music Res. **45**(1), 1–16 (2016)

20. Maas, A.L., Hannun, A.Y., Ng, A.Y.: Rectifier nonlinearities improve neural network acoustic models. In: Proceedings of icml, vol. 30, p. 3 (2013)
21. Meinshausen, N., Bühlmann, P.: Stability selection. J. R. Stat. Soc.: Ser. B (Stat. Methodol.) **72**(4), 417–473 (2010)
22. Meredith, D.: COSIATEC and SIATECCompress: pattern discovery by geometric compression. In: International Society for Music Information Retrieval Conference. International Society for Music Information Retrieval (2013)
23. Meredith, D.: Computational Music Analysis, vol. 62. Springer, Heidelberg (2016). https://doi.org/10.1007/978-3-319-25931-4
24. Mor, N., Wolf, L., Polyak, A., Taigman, Y.: A universal music translation network. arXiv preprint arXiv:1805.07848 (2018)
25. Ren, S., He, K., Girshick, R., Sun, J.: Faster R-CNN: towards real-time object detection with region proposal networks. In: Cortes, C., Lawrence, N.D., Lee, D.D., Sugiyama, M., Garnett, R. (eds.) Advances in Neural Information Processing Systems 28, pp. 91–99. Curran Associates, Inc. (2015). http://papers.nips.cc/paper/5638-faster-r-cnn-towards-real-time-object-detection-with-region-proposal-networks.pdf
26. Rezende, D.J., Mohamed, S., Wierstra, D.: Stochastic backpropagation and approximate inference in deep generative models. arXiv preprint arXiv:1401.4082 (2014)
27. Roberts, A., Engel, J., Raffel, C., Hawthorne, C., Eck, D.: A hierarchical latent vector model for learning long-term structure in music. arXiv preprint arXiv:1803.05428 (2018)
28. Rumelhart, D.E., Hinton, G.E., Williams, R.J.: Learning internal representations by error propagation. Technical report, California University San Diego La Jolla Institute for Cognitive Science (1985)
29. Vincent, P., Larochelle, H., Lajoie, I., Bengio, Y., Manzagol, P.A.: Stacked denoising autoencoders: learning useful representations in a deep network with a local denoising criterion. J. Mach. Learn. Res. **11**(Dec), 3371–3408 (2010)
30. Wang, X., Yu, F., Dou, Z.Y., Gonzalez, J.E.: Skipnet: learning dynamic routing in convolutional networks. arXiv preprint arXiv:1711.09485 (2017)
31. Watson, J., Holmes, C., et al.: Approximate models and robust decisions. Stat. Sci. **31**(4), 465–489 (2016)
32. Witek, M.A., Carlsen, K.: Simultaneous rhythmic events with different schematic affiliations: microtiming and dynamic attending in two contemporary R&B grooves. In: Musical Rhythm in the Age of Digital Reproduction, pp. 51–68. Routledge (2016)
33. Yunpeng, C., Xiaojie, J., Bingyi, K., Jiashi, F., Shuicheng, Y.: Sharing residual units through collective tensor factorization in deep neural networks. arXiv preprint arXiv:1703.02180 (2017)
34. Zeiler, M.D., Krishnan, D., Taylor, G.W., Fergus, R.: Deconvolutional networks (2010)

Assessing the Potential of Classical Q-learning in General Game Playing

Hui Wang$^{(\boxtimes)}$, Michael Emmerich, and Aske Plaat

Leiden Institute of Advanced Computer Science, Leiden University,
Leiden, The Netherlands
h.wang.13@liacs.leidenuniv.nl
http://www.cs.leiden.edu

Abstract. After the recent groundbreaking results of AlphaGo and
AlphaZero, we have seen strong interests in deep reinforcement learning
and artificial general intelligence (AGI) in game playing. However, deep
learning is resource-intensive and the theory is not yet well developed.
For small games, simple classical table-based Q-learning might still be
the algorithm of choice. General Game Playing (GGP) provides a good
testbed for reinforcement learning to research AGI. Q-learning is one
of the canonical reinforcement learning methods, and has been used by
(Banerjee & Stone, IJCAI 2007) in GGP. In this paper we implement
Q-learning in GGP for three small-board games (Tic-Tac-Toe, Connect
Four, Hex), to allow comparison to Banerjee et al. We find that Q-
learning converges to a high win rate in GGP. For the ϵ-greedy strategy, we propose a first enhancement, the dynamic ϵ algorithm. In addition, inspired by (Gelly & Silver, ICML 2007) we combine online search
(Monte Carlo Search) to enhance offline learning, and propose QM-
learning for GGP. Both enhancements improve the performance of classical Q-learning. In this work, GGP allows us to show, if augmented
by appropriate enhancements, that classical table-based Q-learning can
perform well in small games.

Keywords: Reinforcement learning · Q-learning ·
General Game Playing · Monte Carlo Search

1 Introduction

Traditional game playing programs are written to play a single specific game,
such as Chess, or Go. The aim of *General* Game Playing [1] (GGP) is to create
adaptive game playing programs; programs that can play more than one game
well. To this end, GGP uses a so-called Game Description Language (GDL) [2].
GDL-authors write game-descriptions that specify the rules of a game. The challenge for GGP-authors is to write a GGP player that will play any game well.
GGP players should ensure that a wide range of GDL-games can be played
well. Comprehensive tool-suites exist to help researchers write GGP and GDL
programs, and an active research community exists [3–6].

© Springer Nature Switzerland AG 2019
M. Atzmueller and W. Duivesteijn (Eds.): BNAIC 2018, CCIS 1021, pp. 138–150, 2019.
https://doi.org/10.1007/978-3-030-31978-6_11

The GGP model follows the state/action/result paradigm of reinforcement learning [7], a paradigm that has yielded many successful problem solving algorithms. For example, the successes of AlphaGo are based on two reinforcement learning algorithms, Monte Carlo Tree Search (MCTS) [8] and Deep Q-learning (DQN) [9,10]. MCTS, in particular, has been successful in GGP [11]. However, few works analyze the potential of Q-learning for GGP, not to mention DQN. The aim of this paper is to be a basis for further research of DQN for GGP.

Q-learning with deep neural networks requires extensive computational resources. Table-based Q-learning might offer a viable alternative for small games. Therefore, following Banerjee [12], in this paper we address the convergence speed of table-based Q-learning. We use three small two-player zero-sum games: Tic-Tac-Toe, Hex and Connect Four, and table-based Q-learning. We introduce two enhancements: dynamic ϵ, and, borrowing an idea from [13], we create a new version of Q-learning, inserting Monte Carlo Search (MCS) into Q-learning, using online search for offline learning[1].

Our contributions can be summarized as follows:

1. **Dynamic ϵ:** We evaluate the classical Q-learning, finding (1) that Q-learning works and converges in GGP, and (2) that Q-learning with a dynamic ϵ can enhance the performance of TD(λ)[2] baseline with a fixed ϵ [12].
2. **QM-learning:** To further improve performance we enhance classical Q-learning by adding a modest amount of Monte Carlo lookahead (QMPlayer) [14]. This improves the convergence rate of Q-learning, and shows that online search can also improve the offline learning in GGP.

The paper is organized as follows. Section 2 presents related work and recalls basic concepts of GGP and reinforcement learning. Section 3 presents the designs of the QPlayer with fixed and dynamic ϵ and QMPlayer for two-player zero-sum games for GGP to assess the potential of classical Q-learning in detail. Section 4 presents the experimental results. Section 5 concludes the paper and discusses directions for future work.

2   Related Work and Preliminaries

2.1   GGP

A General Game Player must be able to accept formal GDL descriptions of a game and play games effectively without human intervention [4], where the GDL has been defined to describe the game rules [15]. An interpreter program [5] generates legal moves (actions) for a specific board (state). Furthermore, a Game Manager (GM) is at the center of the software ecosystem. The GM interacts with game players through TCP/IP protocol to control the match. The GM manages game descriptions and matches records and temporary states of matches while the game is running. The system also contains a viewer interface for users who are interested in running matches and a monitor to analyze the match process.

---

[1] Source code: https://github.com/wh1992v/ggp-rl.
[2] One of temporal difference methods, see [7].

2.2   Reinforcement Learning

Since Watkins proposed Q-learning in 1989 [16], much progress has been made in reinforcement learning [17,18]. However, few works report on the use of Q-learning in GGP. In [12], Banerjee and Stone propose a method to create a general game player to study knowledge transfer, combining Q-learning and GGP. Their aim is to improve the performance of Q-learning by transferring the knowledge learned in one game to a new, but related, game. They found knowledge transfer with Q-learning to be expensive. In [13], Gelly and Silver combine online and offline knowledge to improve learning performance.

Recently, DeepMind published work on mastering Chess and Shogi by self-play with a deep, generalized reinforcement learning algorithm [19]. With a series of landmark publications from AlphaGo to AlphaZero [10,19,20], these works showcase the promise of general reinforcement learning algorithms. However, such learning algorithms are very resource-intensive and typically require special GPU/TPU hardware. Furthermore, the neural network-based approach is quite inaccessible to theoretical analysis. Therefore, in this paper we study the performance of table-based Q-learning.

In GGP, variants of MCTS [8] are used with great success [11]. Méhat et al. combined UCT (Upper Confidence bound applied to Trees) and nested MCS for single-player general game playing [21]. Cazenave et al. further proposed a nested MCS for two-player games [22]. Monte Carlo techniques have proved a viable approach for searching intractable game spaces and other optimization problems [23]. Therefore, in this paper we combine MCS to improve performance.

2.3   Q-learning

A basic distinction between reinforcement learning methods is that of "on-policy" and "off-policy" methods. On-policy methods attempt to evaluate or improve the policy that is used to make decisions, whereas off-policy methods evaluate or improve a policy *different* from that used to make decisions [7]. Q-learning is an off-policy method. The reinforcement learning model consists of an *agent*, a set of states S, and a set of actions A available in state S [7]. The agent can move to the next state s', $s' \in S$ from state s after following action a, $a \in A$, denoted as $s \xrightarrow{a} s'$. After finishing the action a, the agent gets an immediate reward $R(s, a)$, usually a numerical score. The cumulative return of current state s by taking the action a, denoted as $Q(s, a)$, is a weighted sum, calculated by $R(s, a)$ and the maximum $Q(s', a')$ value of all next states:

$$Q(s, a) = R(s, a) + \gamma \, max_{a'} Q(s', a') \qquad (1)$$

where $a' \in A'$ and A' is the set of actions available in state s'. γ is the discount factor of $max_{a'} Q(s', a')$ for next state s'. $Q(s, a)$ can be updated by online interactions with the environment using the following rule:

$$Q(s, a) \leftarrow (1 - \alpha) \, Q(s, a) + \alpha \, (R(s, a) + \gamma \, max_{a'} Q(s', a')) \qquad (2)$$

where $\alpha \in [0, 1]$ is the learning rate. The Q-values are guaranteed to converge after iteratively updating.

3   Design

3.1   Classical Q-learning for Two-Player Games

GGP games in our experiments are two-player zero-sum games that alternate moves. Therefore, we can use the same rule, see Algorithm 1 line 5, to create $R(s, a)$, rather than to use a reward table. In our experiments, we set $R(s, a) = 0$ for non-terminal states, and call the $getGoal()$ function for terminal states. In order to improve the learning effectiveness, we update the $Q(s, a)$ table only at the end of the match. During offline learning, QPlayer uses an ϵ-greedy strategy to balance exploration and exploitation towards convergence. While the ϵ-greedy strategy is enabled, QPlayer will perform a random action. Otherwise, QPlayer will perform the best action according to Q(S,A) table. If no record matches current state, QPlayer will perform a random action. The pseudo code for this algorithm is given in Algorithm 1.

Algorithm 1. Classical Q-learning Player with Static ϵ

1: **function** QPLAYER(current state s, learning rate α, discount factor γ, Q table: $Q(S, A)$)
2:     **for** each match **do**
3:         **if** s terminates **then**
4:             **for** each (s, a) from end to the start in current match record **do**
5:                 R(s,a) $=s'$ is terminal state? getGoal(s', myrole) : 0
6:                 Update $Q(s, a) \leftarrow (1 - \alpha) Q(s, a) + \alpha (R(s, a) + \gamma \, max_{a'} Q(s', a'))$
7:         **else**
8:             **if** ϵ-greedy is enabled **then**
9:                 selected_action = Random()
10:            **else**
11:                selected_action = SelectFromQTable()
12:                **if** no s record in $Q(S, A)$ **then**
13:                    ***selected_action = Random()***
14:   ▷ To be changed for different versions
15:            performAction(s, selected_action)
16:     **return** $Q(S, A)$

3.2   Dynamic ϵ Enhancement

In contrast to the baseline of [12], which uses a fixed ϵ value, we use a dynamically decreasing ϵ-greedy Q-learning [17]. In our implementation, we use the function

$$\epsilon(m) = \begin{cases} a(\cos(\frac{m}{2l}\pi)) + b & m \le l \\ 0 & m > l \end{cases} \tag{3}$$

for ϵ, where m is the current match count, and l is a number of matches we set in advance to control the decaying speed of ϵ. During offline learning, if $m = l$, ϵ

decreases to 0. a and b is set to limit the range of ϵ, where $\epsilon \in [b, a+b]$, $a, b \geq 0$ and $a + b \leq 1$. The player generates a random number num where $num \in [0, 1]$. If $num < \epsilon$, the player will explore a random action, else the player will exploit best action from the currently learnt $Q(s, a)$ table. Note that in this function, in order to assess the potential of Q-learning in detail, we introduce l for controlling the decay of ϵ. This parameter determines the value and changing speed of ϵ in current match count m. Instances in our experiments are shown in Fig. 1:

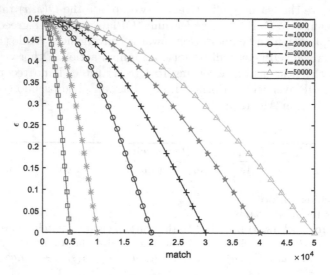

Fig. 1. Decaying Curves of ϵ with Different l. Every curve decays from 0.5 (learning start, explore & exploit) to 0 ($m \geq l$, fully exploit).

3.3   QM-learning Enhancement

The main idea of Monte Carlo Search [14] is to make some lookahead probes from a non-terminal state to the end of the game by selecting random moves for the players to estimate the value of that state. To apply Monte Carlo in game playing, we use a time-limited version, since in competitive game playing time for each move is an important factor for the player to consider. The time limited MCS in GGP that we use is written as ***MonteCarloSearch(time_limit)***.

In Algorithm 1 (line 13), we see that a *random action* is chosen when QPlayer can not find an existing value in the $Q(s, a)$ table. In this case, QPlayer acts like a random player, which will lead to a low win rate and slow learning speed. In order to address this problem, we introduce a variant of Q-learning combined with MCS. MCS performs a time limited lookahead to find better moves. The more time it has, the better the action it finds will be. To achieve this, we use ***selected_action = MonteCarloSearch(time_limit)*** to replace the line 13, giving QM-learning. By adding MCS, we effectively add a local version of the last two stages of MCTS to Q-learning: the playout and backup stage [8].

4 Experiments and Results

4.1 Dynamic ϵ Enhancement

We create ϵ-greedy Q-learning players ($\alpha = 0.1$, $\gamma = 0.9$) with fixed $\epsilon = 0.1$, 0.2 and with dynamically decreasing $\epsilon \in [0, 0.5]$ to play 30000 matches first ($l = 30000$) against a Random player, respectively. During these 30000 matches, the dynamic ϵ decreases from 0.5 to 0 based on the decay function, see Eq. 3. The fixed values for ϵ are 0.1 and 0.2, respectively. After 30000 matches, fixed ϵ is also set to 0 to continue the competition. For Tic-Tac-Toe, results in Fig. 2 show that dynamically decreasing ϵ performs better. We see that the final win rate of dynamically decreasing ϵ is 4% higher than fixed $\epsilon = 0.1$ and 7% higher than fixed $\epsilon = 0.2$. Therefore, in the rest of the experiments, we use dynamic ϵ for further improvements.

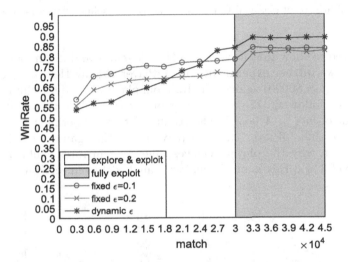

Fig. 2. Win rate of the fixed and dynamic ϵ Q-learning Player vs a Random Player Baseline. In the white part, the player uses ϵ-greedy to learn; in the grey part, all players set $\epsilon = 0$ (stable performance). The color code of the rest figures are the same

To enable comparison with previous work, we implemented TD(λ), the baseline learner of [12] ($\alpha = 0.3$, $\gamma = 1.0$, $\lambda = 0.7$, $\epsilon = 0.01$), and dynamic ϵ learner($\alpha = 0.1$, $\gamma = 0.9$, $\epsilon \in [0, 0.5]$, $l = 30000$, Algorithm 1). For Tic-Tac-Toe, from Fig. 3, we find that although the TD(λ) player converges more quickly initially (win rate stays at about 75.5% after 9000th match) our dynamic ϵ player performs better when the value of ϵ decreases dynamically with the learning process.

Experiments above suggest the following conclusions: that (1) classical Q-learning is applicable to a GGP system, and that (2) a dynamic ϵ can enhance the performance of fixed ϵ. However, beyond the basic applicability in a single

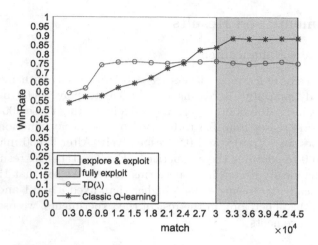

Fig. 3. Win rate of classical Q-learning and [11] Baseline Player vs Random.

game, we need to show that it can do so (1) *efficiently*, and (2) in more than one game. Thus, we further experiment with QPlayer to play Hex ($l = 50000$) and Connect Four ($l = 80000$) against the Random player. In order to limit excessive learning times, following [12], we play Hex on a very small 3×3 board, and play ConnectFour on a 4×4 board. The results of these experiments are given in Fig. 4. We see that QPlayer can also play these other games effectively. Note that the reason why the player achieves different win rates could be that the game space of 3×3 Hex is much smaller than 4×4 ConnectFour.

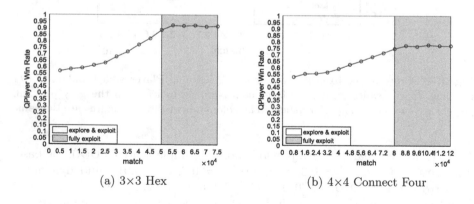

(a) 3×3 Hex

(b) 4×4 Connect Four

Fig. 4. Win rate of QPlayer vs Random Player in different games. For Hex and Connect-Four the win rate of Q-learning also converges

However, so far, all our games are small. QPlayer should be able to learn to play larger games. The complexity influences how many matches the QPlayer should learn. We will now show results to demonstrate how QPlayer performs while playing more complex games. We make QPlayer play Tic-Tac-Toe (a line of 3 stones is a win, $l = 50000$) in 3×3, 4×4 and 5×5 boards, respectively, and show the results in Fig. 5.

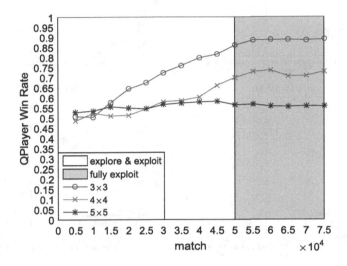

Fig. 5. Win rate of QPlayer vs Random in Tic-Tac-Toe on different board size. For larger board sizes convergence slows down

The results show that with the increase of game board size, QPlayer performs worse. For larger boards can not achieve convergence. The reason for the lack of convergence is that QPlayer has not learned enough knowledge. Our experiments also show that for table-based Q-learning in GGP, large game complexity leads to slow convergence, which confirms the well-known drawback of classical Q-learning.

4.2   QM-learning Enhancement

The second contribution of this paper is QM-learning enhancement, we implement the QPlayer and QMPlayer based on Algorithm 1 and Sect. 3.3. For both players, we set parameters to $\alpha = 0.1$, $\gamma = 0.9$, $\epsilon \in [0, 0.5]$ respectively and we set the $l = 5000, 10000, 20000, 30000, 40000, 50000$, respectively. For QMPlayer, we set $time\_limit = 50$ ms. Next we make them play the game with the Random baseline player for $1.5 \times l$ matches for 5 rounds respectively. The comparison between QPlayer and QMPlayer is shown in Fig. 6.

Figure 6(a) shows that QPlayer has the most unstable performance (the largest variance in 5 experiments) and only wins around 55% matches after training 5000 matches. Figure 6(b) illustrates that after training 10000 matches QPlayer wins about 80% matches. However, during the exploration period (the

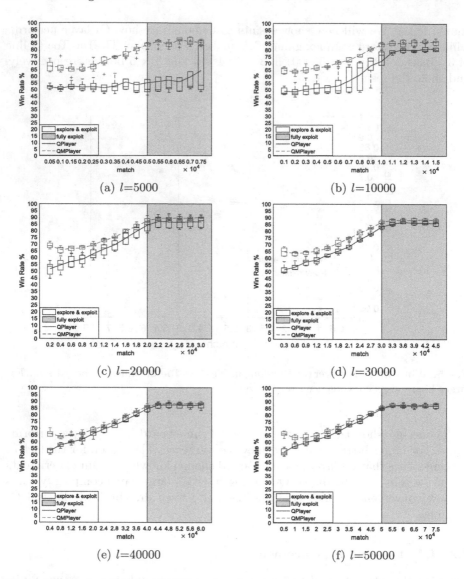

Fig. 6. Win rate of QMPlayer (QPlayer) vs Random in Tic-Tac-Toe for 5 experiments. Small Monte Carlo lookaheads improve the convergence of Q-learning, especially in the early part of learning. QMPlayer always outperforms Qplayer

white part of the figure) the performance is still very unstable. Figure 6(c) shows that QPlayer wins about 86% of the matches while learning 20000 matches still with high variance. Figure 6(d), (e), (f), show us that after training 30000, 40000, 50000 matches, QPlayer gets a similar win rate, which is nearly 86.5% with smaller and smaller variance.

In Fig. 6(a), QMPlayer gets a high win rate (about 67%) at the very beginning. Then the win rate decreases to 66% and 65%, and then increases from 65% to around 84% at the 5000th match. Finally, the win rate stays at around 85%. Also in the other sub figures, for QMPlayer, the curves all decrease first and then increase until reaching a stable state. This is because at the very beginning, QMPlayer chooses more actions from MCS. Then as the learning period moves forward, it chooses more actions from Q table.

Overall, as the l increases, the win rate of QPlayer becomes higher until leveling off around 86.5%. The variance becomes smaller and smaller, which proves that Q-learning can achieve convergence in GGP games and that a proper ϵ decaying speed makes sense for classical Q-learning. Note that in every sub figure, QMPlayer can always achieve a higher win rate than QPlayer, not only at the beginning but also at the end of the learning period. Overall, QMPlayer achieves a better performance than QPlayer with the higher convergence win rate (at least 87.5% after training 50000 matches). To compare the convergence speeds of QPlayer and QMPlayer, we summarize the convergence win rates of different l according to Fig. 6 in Fig. 7.

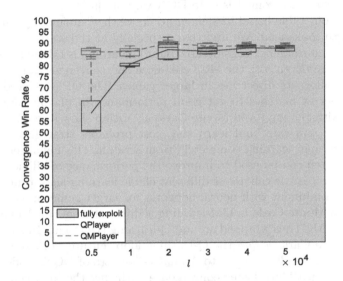

Fig. 7. Convergence win rate of QMPlayer (QPlayer) vs Random in Tic-Tac-Toe

These results show that combining online MCS with classical Q-learning for GGP can improve the win rate both at the beginning and at the end of the offline learning period. The main reason is that QM-learning allows the $Q(s, a)$ table to be filled quickly with good actions from MCS, achieving a quick and direct learning rate. It is worth to note that, QMPlayer will spend slightly more time (at most is *search time limit* × *number of (state-action) pairs*) in training than QPlayer. It will be time consuming for MCS to compute a large game, and this is also the essential drawback of table-based Q-learning, so currently QM-learning is also only applicable for small games.

5   Conclusion

This paper examines the applicability of Q-learning, a canonical reinforcement learning algorithm, to create general players for GGP programs. Firstly, we show how good canonical implementations of Q-learning perform on GGP games. The GGP system allows us to easily use three real games for our experiments: Tic-Tac-Toe, Connect Four, and Hex. We find that (1) Q-learning is indeed general enough to achieve convergence in GGP games. However, we also find that convergence is slow. In accordance with Banerjee [12], who used a static value for ϵ, we find that (2) a value for ϵ that changes with the learning phases gives better performance (start with more exploration, become more greedy later on). The table-based implementation of Q-learning facilitates theoretical analysis, and comparison against some baselines [12]. However, it is only suitable for small games. A neural network implementation facilitates the study of larger games, and allows meaningful comparison to DQN variants [9].

Still using our table-based implementation, we then enhance Q-learning with an MCS based lookahead. We find that, especially at the start of the learning, this speeds up convergence considerably. Our Q-learning is table-based, limiting it to small games. Even with the MCS enhancement, convergence of QM-learning does not yet allow its direct use in larger games. The QPlayer needs to learn a large number of matches to get good performance in playing larger games. The results with the improved Monte Carlo algorithm show a real improvement of the player's win rate, and learn the most probable strategies to get high rewards faster than learning completely from scratch. This enhancement shows how online search can be used to improve the performance of offline learning in GGP. On this basis, we can assess different offline learning algorithms (or follow Gelly [13] to combine it with neural networks for larger games in GGP).

Our use of Monte Carlo in QM-learning is different from the AlphaGo architecture, where MCTS is wrapped around Q-learning (DQN) [9]. In our approach, we insert Monte Carlo *within* the Q-learning loop. Future work should show if our QM-learning results transfer to AlphaGo-like uses of DQN inside MCTS, if QM-learning can achieve faster convergence, reducing the high computational demands of AlphaGo [19]. Additionally, we plan to study nested MCS in Q-learning [22]. Implementing Neural Network based players also allows the study of more complex GGP games.

Acknowledgments. Hui Wang acknowledges financial support from the China Scholarship Council (CSC), CSC No. 201706990015.

References

1. Genesereth, M., Love, N., Pell, B.: General game playing: overview of the AAAI competition. AI Mag. **26**(2), 62–72 (2005)
2. Love, N., Hinrichs, T., Haley, D., Schkufza, E., Genesereth, M.: General game playing: game description language specification. Stanford Technical report LG-2006-1 (2008)
3. Kaiser, D.M.: The design and implementation of a successful general game playing agent. In: International Florida Artificial Intelligence Research Society Conference, pp. 110–115. AAAI Press, California (2007)
4. Genesereth, M., Thielscher, M.: General game playing. Synth. Lect. Artif. Intell. Mach. Learn. **8**(2), 1–229 (2014)
5. Świechowski, M., Mańdziuk, J.: Fast interpreter for logical reasoning in general game playing. J. Logic Comput. **26**(5), 1697–1727 (2014)
6. Wang, H., Tang, Y., Liu, J., Chen, W.: A search optimization method for rule learning in board games. In: Geng, X., Kang, B.-H. (eds.) PRICAI 2018. LNCS (LNAI), vol. 11013, pp. 174–181. Springer, Cham (2018). https://doi.org/10.1007/978-3-319-97310-4_20
7. Sutton, R.S., Barto, A.G.: Reinforcement Learning: An Introduction, 2nd edn. MIT Press, Cambridge (1998)
8. Browne, C.B., Powley, E., Whitehouse, D., et al.: A survey of Monte Carlo tree search methods. IEEE Trans. Comput. Intell. AI Games **4**(1), 1–43 (2012)
9. Mnih, V., Kavukcuoglu, K., Silver, D., et al.: Human-level control through deep reinforcement learning. Nature **518**(7540), 529–533 (2015)
10. Silver, D., Huang, A., Maddison, C.J., et al.: Mastering the game of Go with deep neural networks and tree search. Nature **529**(7587), 484–489 (2016)
11. Mehat, J., Cazenave, T.: Monte-Carlo tree search for general game playing. Univ. Paris **8** (2008)
12. Banerjee, B., Stone, P.: General game learning using knowledge transfer. In: Veloso, M.M. (ed.) International Joint Conference on Artificial Intelligence 2007, pp. 672–677 (2007)
13. Gelly, S., Silver, D.: Combining online and offline knowledge in UCT. In: Proceedings of the 24th International Conference on Machine Learning, pp. 273–280 (2007)
14. Robert, C.P.: Monte Carlo Methods. Wiley, Hoboken (2004)
15. Thielscher, M.: The general game playing description language is universal. In: Toby Walsh. International Joint Conference on Artificial Intelligence 2011, vol. 22, no. 1, pp. 1107–1112. AAAI Press, California (2011)
16. Watkins, C.J.C.H.: Learning from Delayed Rewards. King's College, Cambridge (1989)
17. Even-Dar, E., Mansour, Y.: Convergence of optimistic and incremental Q-learning. In: Dietterich, T.G., Becker, S., Ghahramani, Z. (eds.) Advances in Neural Information Processing Systems 2001, pp. 1499–1506. MIT Press, Cambridge (2001)
18. Hu, J., Wellman, M.P.: Nash Q-learning for general-sum stochastic games. J. Mach. Learn. Res. **4**, 1039–1069 (2003)
19. Silver, D., Hubert, T., Schrittwieser, J., et al.: Mastering Chess and Shogi by self-play with a general reinforcement learning algorithm. arXiv preprint arXiv:1712.01815 (2017)
20. Silver, D., Schrittwieser, J., Simonyan, K., et al.: Mastering the game of go without human knowledge. Nature **550**(7676), 354–359 (2017)

21. Méhat, J., Cazenave, T.: Combining UCT and nested Monte Carlo search for single-player general game playing. IEEE Trans. Comput. Intell. AI Games **2**(4), 271–277 (2010)
22. Cazenave, T., Saffidine, A., Schofield, M.J., Thielscher, M.: Nested Monte Carlo search for two-player games. In: Schuurmans, D., Wellman, M.P. (eds.) AAAI Conference on Artificial Intelligence 2016, vol. 16, pp. 687–693. AAAI Press, California (2016)
23. Ruijl, B., Vermaseren, J., Plaat, A., Herik, J.: Combining simulated annealing and Monte Carlo tree search for expression simplification. In: Duval, B., Jaap van den Herik, H., Loiseau, S., Filipe, J. (eds.) Proceedings of the 6th International Conference on Agents and Artificial Intelligence 2014, vol. 1, pp. 724–731. SciTePress, Setúbal, Portugal (2014)

Visual Rationalizations in Deep Reinforcement Learning for Atari Games

Laurens Weitkamp[1(✉)], Elise van der Pol[2], and Zeynep Akata[2]

[1] Informatics Institute, University of Amsterdam, Amsterdam, The Netherlands
laurens.weitkamp@student.uva.nl
[2] UvA-Bosch Delta Lab, University of Amsterdam, Amsterdam, The Netherlands

Abstract. Due to the capability of deep learning to perform well in high dimensional problems, deep reinforcement learning agents perform well in challenging tasks such as Atari 2600 games. However, clearly explaining why a certain action is taken by the agent can be as important as the decision itself. Deep reinforcement learning models, as other deep learning models, tend to be opaque in their decision-making process. In this work, we propose to make deep reinforcement learning more transparent by visualizing the evidence on which the agent bases its decision. In this work, we emphasize the importance of producing a justification for an observed action, which could be applied to a black-box decision agent.

Keywords: Explainable AI · Reinforcement learning · Deep learning

1 Introduction

Due to strong results on challenging benchmarks over the last few years, enabled by the use of deep neural networks as function approximators [6,11,17] deep reinforcement learning has become an increasingly active field of research. While neural networks allow reinforcement learning methods to scale to complex problems with large state spaces, their decision-making is opaque and they can fail in non-obvious ways, for example, if the network fails to generalize well and chooses an action based on the wrong feature. Moreover, recent work [7] has shown that these methods can lack robustness, with large differences in performance when varying hyperparameters, deep learning libraries or even the random seed. Gaining insight into the decision-making process of reinforcement learning (RL) agents can provide new intuitions about why and how they fail. Moreover, agents that can justify with visual elements why a prediction is consistent to a user are more likely to be trusted [18]. Generating such post-hoc explanations, also referred to as rationalizations, does not only increase trust, but also it is a key component for understanding and interacting with them [3]. Motivated by explainability as a means to make the black-box neural networks transparent, we propose to visualize the decision process of a reinforcement learning agent by using Grad-CAM [15].

© Springer Nature Switzerland AG 2019
M. Atzmueller and W. Duivesteijn (Eds.): BNAIC 2018, CCIS 1021, pp. 151–165, 2019.
https://doi.org/10.1007/978-3-030-31978-6_12

Grad-CAM creates an activation map that shows prominent spaces of activation given an input image and class, typically in an image classification task. The activation map is calculated through a combination of the convolutional neural network weights and the gradient activations created during a forward pass of the input image and class in the neural network.

Applying this method instead to a reinforcement learning agent, it wil be used to construct action-specific activation maps that highlight the regions of the image that are the most important evidences, for the predicted action of the RL agent. We evaluate these visualizations on three Atari 2600 games using the OpenAI Gym wrapper, created precisely to tackle difficult problems in deep reinforcement learning. The range of games in the wrapper are diverse in difficulty: they have different long-term reward mechanics and a different action space per game. These difficulties are of interest when looking to explain why the agent takes a specific action given a state.

This paper is structured as follows: The next section, Sect. 2, discusses related works in both reinforcement learning and explainable AI. Section 3 presents the visual rationalization model and explains how it is adapted to reinforcement learning tasks. Following after that is Sect. 4 which provides the setup required for experiments. This section also provides the results for the rationalization model, including where the model fails. The last section, Sect. 5, provides an conclusion to the experiments.

2   Related Work

In this section, we discuss previous works relevant to reinforcement learning and explainable artificial intelligence.

2.1   Deep Reinforcement Learning

In general, there are two main methods in deep reinforcement learning. The first method uses a neural network to approximate the value function that estimates the value of state, action pairs as to infer a policy. One such value function estimation model is called the Deep-Q Network (DQN), which has had garnered much attention to the field of deep reinforcement learning due to impressive results on challenging benchmarks such as Atari 2600 games [11]. Since the release of this model, a range of modifications have been proposed that have improved this model such as the Deep Recurrent-Q model, the Double DQN model and the Rainbow DQN model [5,9,19].

The second method used in deep reinforcement learning approximates the policy directly, by parameterizing the policy and using the gradient of these parameters to calculate an optimal policy. This method is called the policy gradient method, and a much cited example of such a method is known as the REINFORCE line of algorithms [20]. More recent examples of policy gradient methods include Trust Region Policy Optimization and Proximal Policy Optimization [13,14].

A hybrid that combines value function methods and policy gradient methods is known as the actor-critic method. In this method, the actor is trying to infer a policy using a state, action pair and the critic is assigning a value to the current state of the actor. In this paper we use the Asynchronous Advantage Actor-Critic (A3C) model which has been used to achieve human level performance on a wide range of Atari 2600 games [10].

2.2 Explainable AI

Generating visual or textual explanations of deep network predictions is a research direction that has recently gained much interest [1,8,12,23]. Following the convention described by Park et al. in [12], we focus on post-hoc explanations, namely rationalizations where a deep network is trained to explain a black box decision maker which is useful in increasing trust for the end user.

Textual rationalizations are explored in Hendricks et al. [8] which proposes a loss function based on sampling and reinforcement learning that learns to generate sentences that realize a global sentence property, such as class specificity. Andreas et al. [1] composes collections of jointly-trained neural modules into deep networks for question answering by decomposing questions into their linguistic substructures, and using these structures to dynamically instantiate modular networks with reusable components.

As for visual rationalizations, Zintgraf et al. [23] propose to apply prediction difference analysis to a specific input. [12] utilizes a visual attention module that justifies the predictions of deep networks for visual question answering and activity recognition. In [4] the authors propose to use a perturbation method that selectively blurs regions to calculate the impact on an RL agent's policy. Although this method demonstrates important regions for the agent's decision making, the method used in this paper highlights important regions without the need for such a perturbation method.

Grad-CAM [15] uses the gradients of any target concept, i.e. predicted action, flowing into the final convolutional layer to produce a coarse localization map highlighting the important regions in the image for predicting the concept. It has been demonstrated on image classification and captioning. In this work, we adapt it to two reinforcement learning tasks to visually rationalize the predicted action.

3 Visual Rationalization Model

In reinforcement learning, an agent interacting with an environment over a series of discrete time steps observes a state[1] $s_t \in \mathcal{S}$, takes an action $a_t \in \mathcal{A}$ and receives a reward r_t and observes the next state $s_{t+1} \in \mathcal{S}$. The agent is tasked

---

[1] Here we assume problems where partial observability can be addressed by representing a state as a small number of past observations.

with finding a policy $\pi : S \times A \to [0, 1]$, a function mapping states and actions to probabilities whose goal is to maximize the discounted sum of rewards:

$$R_t = \sum_{k=0}^{\infty} \gamma^k r_{t+k+1} \tag{1}$$

which is the return with discount factor $\gamma \in [0, 1]$.

3.1 Asynchronous Advantage Actor Critic Learning

Gradient based actor-critic methods split the agent in two components: an actor that interacts with the environment using a policy $\pi(a|s; \theta)$, and a critic that assigns values to these actions using the value function $V(s; \theta)$. Both the policy and the value function are directly parameterized by θ. Updating the policy and value function is done through gradient descent

$$\theta_{t+1} = \theta_t + \nabla_{\theta_t} \log \pi(a_t|s_t; \theta_t) A_t. \tag{2}$$

With $A_t = R_t - V(s_t; \theta_t)$, an estimation of the advantage function [21]. In [10], the policy gradient actor-critic uses a series of asynchronous actors that all send policy-gradient updates to a single model that keeps track of the parameters θ. In our implementation the actor output is a softmax vector of size $|A|$, the total number of actions the agent can take in the specific environment. Because our visual rationalization model uses the actor output only, the scalar critic output will be ignored for the purposed of this paper. However, in future work, exploring the critic's explanations could be of interest. To ensure exploration early on an entropy regularization term H is introduced with respect to the policy gradient,

$$\theta_{t+1} = \theta_t + \nabla_{\theta_t} \log \pi(a_t|s_t; \theta_t) A_t + \beta \nabla_{\theta_t} H(\pi(s_t; \theta_t)), \tag{3}$$

where β is a hyper parameter discounting the entropy regularization.

3.2 Visual Rationalization

Our visual rationalization is based on Grad-CAM [15], and constitutes of computing a class-discriminative localization map $L_{GradCAM}^s \in R^{u \times v}$ using the gradient of any target class. These gradients are global-average-pooled to obtain the neuron importance weights a_k^c for class c, for activation layer k in the CNN[2]:

$$\alpha_k^c = \frac{1}{Z} \sum_i \sum_j \frac{\partial y^c}{\partial L_{ij}^k}. \tag{4}$$

---

[2] k is usually chosen to be the last convolutional layer in the CNN.

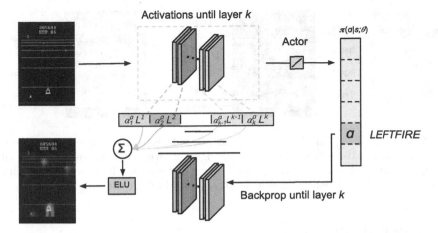

Fig. 1. The model takes as input a state, calculates the state-action $\pi(a|s;\theta)$ policy and then produces a gradient-based activation map based on the state, action pair. This activation map can then be overlayed on the original state to indicate evidence that the agent has to take the action. In this Figure, the agent chooses to take the action LEFTFIRE which would make the agent go one step to the left and then shoot up. The activation map is highlighting the agent (bottom), incoming debris (upper-right) and an incoming enemy (upper-mid).

Adapting this method in particular to the A3C actor output, let h^a be the score for action a before the softmax, α_k^a now represents the importance weight for state a in activation layer k:

$$\alpha_k^a = \frac{1}{Z}\sum_i\sum_j \frac{\partial h^a}{\partial L_{ij}^k},\tag{5}$$

with $|h| = |\mathcal{A}|$, the total amount of actions the agent can take. The gradient then gets weighted by the forward-pass activations L^k and passes an ELU activation[3] to produce a weighted class activation map:

$$L_{GradCAM}^a = ELU(\sum_{k=1}^{K} \alpha_k^a L^k).\tag{6}$$

This activation map has values in the range $[0, 1]$ with higher weights corresponding to a stronger response to the input state. This can be applied to the critic output in the same fashion. The resulting activation map can bilinearly extrapolated to the size of the input state and can then be overlayed on top of this state to produce a high-quality heatmap that indicate regions that motivate the agent to take action a. A visual representation of this process is depicted in Fig. 1.

---

[3] the Exponential Linear Unit has been chosen in favor of the ReLU used in the original Grad-CAM paper due to the dying ReLU effect described in [22].

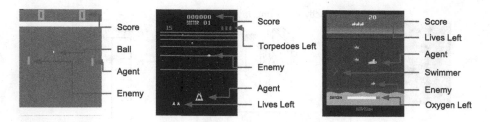

Fig. 2. A detailed explanation of the Pong, BeamRider and Seaquest game frames, respectively from Atari 2600 games [2]. The agent is situated in an environment with (multiple) moving enemies, other moving objects and semi-static objects (for example the torpedoes left in BeamRider and the oxygen bar in Seaquest).

4 Experiments

In this section, we first provide the details of our experimentation setup. We then show qualitative examples evaluating how our model performs in three of the Atari games. Throughout this section, red bounding boxes and red arrows indicate important regions of a state.

4.1 Setup

The Atari 2600 game environment is provided through the Arcade Learning Environment wrapper in the OpenAI Gym framework [2]. The framework has multiple version of each game but for the purpose of this paper the NoFrameskip-v4 environment will be used (OpenAI considers NoFrameskip the canonical Atari environment in gym and v4 is the latest version). Each state is represented as a $210 \times 160 \times 3$ pixel image with a 128-colour palette, and each state is preprocessed to a $84 \times 84 \times 1$ image as input to the network. A side-effect of this preprocessing is that the visual score will be removed from the state in most games, but the agent still gets the reward per state implicitly through the environment.

In our experiments, we use three Atari games, namely Pong, BeamRider and Seaquest, all depicted in Fig. 2. All three games have a different action space (see Table 1), and a different long term-reward system for the agent to learn.

Pong. Pong has six actions with three of the six being redundant (FIRE is equal to NOOP, LEFT is equal to LEFTFIRE and RIGHT is equal to RIGHTFIRE). The agent is displayed on the right and the enemy on the left and the first player to score 21 goals wins.

Table 1. Action space of Pong, BeamRider and Seaquest in the Atari 2600 OpenAI wrapper. Each agent from top to bottom has an increasing amount of actions.

	NOOP	FIRE	UP	LEFT	RIGHT	DOWN	LEFT FIRE	RIGHT FIRE	UP FIRE	UP LEFT	UP RIGHT	DOWN FIRE	DOWN LEFT	DOWN RIGHT	UP LEFT FIRE	UP RIGHT FIRE	DOWN LEFT FIRE	DOWN RIGHT FIRE
Pong	x	x		x	x		x	x										
BeamRider	x	x	x	x	x		x	x	x	x								
Seaquest	x	x	x	x	x	x	x	x	x	x	x	x	x	x	x	x	x	x

BeamRider. In BeamRider the agent is displayed at the bottom and the agent has to traverse a series of sectors where each sector contains 15 enemies (remaining enemies is displayed at the top-left) and a boss at the end. The agent has three torpedoes that can be used specifically to kill the sector boss, but these can also be used to destroy debris that appear in later sectors. Learning how to use the torpedoes correctly is not necessary to succeed in the game, but it provides for long-term rewards in the form of bonus points.

Seaquest. In Seaquest the agent is dependent on a limited amount of oxygen, depicted at the bottom of the state. The agent can ascend to the surface which will refill the oxygen bar and it drops off any swimmers that the agent has picked up along the way for a bonus reward. Resurfacing requires learning a long-term reward dependency which is not easily learned [16]. Surfacing is not just used to refill the oxygen bar but also to drop-off any swimmer that the agent has found underwater which results in additional points. A different way to get a positive reward is to kill sharks.

4.2   Learning A Policy

Training an agent to gain human-like or superhuman-like performance in a complex environment can take millions of input frames. In this section we take the same approach as Greydanus et al. [4], in which the authors argue that deep RL agents, during training, discard policies in favor for better ones. Seeing how an agent is reacting to different situations at different times of training might make it clear how an agent is trying to maximize long-term rewards. To demonstrate this, two agents have been trained for a different number of frames. The first model which will be called the *Full Agent* has been trained using (at least) 40 million frames. The second agent which will be called the *Half Agent* has been trained using 20 million frames, except for the case of Pong where it has been trained using 500,000 frames, due to the fact that Pong is an easier game to learn. The mean score and variance can be found in Table 2. For both games a sequence

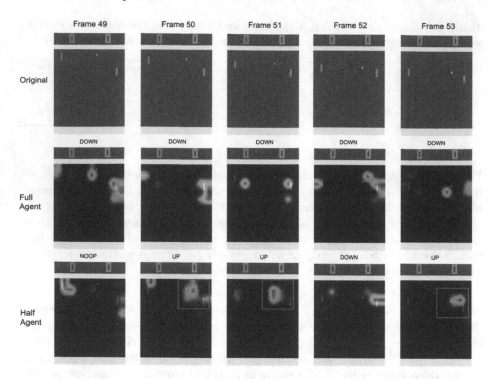

Fig. 3. Manually sampled states from the game Pong, combined with the Full Agent and the Half Agent's actions Grad-CAM outputs based on these states. Indicated in the red boxes is the tracking behavior exhibited by the Half Agent. Best viewed in high resolution in color. (Color figure online)

of states were manually sampled, after which both agents have evaluated[4] the state to learn spatial-temporal information. States were manually sampled by having a person (one of the authors) play one episode of each game. The states were sampled manually to make sure the samples were not biased towards one agents' policy.

Pong. For Pong, the Full Agent has learned to shoot the ball in such a way that it scores by hitting the ball only once each round. The initial round might differ, but after that all rounds are the same: the Full Agent shoots the ball up high which makes the ball bounce off the wall all the way down over the opponent's side, at which point the agent retreats to the lower right corner. This would indicate that the Full Agent is not reacting to the ball most of the time, but is waiting to exploit a working strategy that allows it to win each round. In contrast, he Half Agent is actively tracking the ball at each step and could

---

[4] *evaluated* in this case means having forwarded each state that has been manually sampled through the model.

Table 2. The mean and variance of both the Full Agent (trained on at least 40 million frames) and the Half Agent (500,000 frames in Pong and 20 million frames in BeamRider) after playing 100 episodes using a greedy strategy. Seaquest's Half Agent is omitted because the Full Agent could not learn how to surface for water.

	Full Agent mean	Full Agent variance	Half Agent mean	Half Agent variance
Pong	21.00	0.00	14.99	0.09
BeamRider	4659.04	1932.58	1597.40	1202.00
Seaquest	1749.00	11.44	N/A	N/A

potentially be losing some rounds because of this. The tracking behavior of the Half Agent is demonstrated in Fig. 3 at frames 50, 51 and 53 indicated with a red box. In these frames the Half Agent's attention is focused on the ball and the corresponding action is to go up to match it.

BeamRider. For BeamRider, both agents have learned to hit enemies but the Full Agent has a higher average return. Looking at Fig. 4, both agents have a measure of attention on the two white enemy saucers, but the intensity of attention differs; the Full Agent has high attention on the enemies, in comparison with the Half Agent which has low attention on the enemies. The Half Agent is either going right which is essentially a NOOP in that area or it could be shooting at the incoming enemy. More interesting are the last two frames: 175 and 176. The attention of the Full Agent turns from the directly approaching enemy saucer to the enemy saucer on the left of it, and the agent would try to move into its direction (LEFTFIRE). The Full Agent's attention in frame 176 is placed in a medium degree at the trajectory of its own laser that will hit the enemy saucer in the next frame. This could indicate that the Full Agent knows it will hit the target and is thus moving away from it, to focus on the other remaining enemy.

From the analysis of both agents another interesting result is discovered: the agents do not learn to *properly* use the torpedoes. At the beginning of each episode/level both agents would fire torpedoes until they are all used up and then continue on as usual. In Fig. 5 this phenomena is demonstrated through a manually sampled configuration evaluated by the Full Agent only (the results are the same for the Half Agent). The torpedoes have not been used yet, on purpose, and there are enemies coming towards the agent at different time-steps. Looking at the Grad-CAM attention map, it would appear to be highly focused on the remaining three torpedoes in the upper right corner indicated by a red box. This occurs even when the action chosen by the agent is not of the UP-variety which would trigger firing a torpedo.

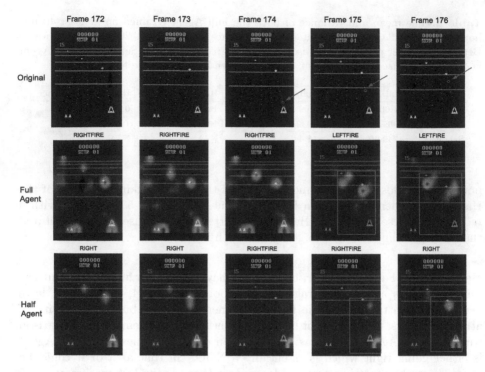

Fig. 4. Manually sampled states from the game BeamRider, combined with the Full Agent and the Half Agent's actions Grad-CAM outputs based on these states. The red boxes indicate the difference in focus of the agents and the arrow indicates the shot fired by the agent. Best viewed in high resolution in color. (Color figure online)

4.3   Agent Failing

A different way of looking at how rationalizations aid in understanding the behavior of an agent is by looking at when an agent fails at its task. In the context of BeamRider and Seaquest, this means looking at the last couple of frames before the agent dies.

BeamRider. In the situation depicted in Fig. 6 the agent is approached by a number of different enemies, one of which only appears after sector 7: the green bounce craft, depicted inside a red box in the first four frames. This is an enemy that can only be destroyed by shooting a torpedo at it, and it jumps from beam to beam trying to hit the agent which is what kills the agent eventually in the last frame. In all frames the Grad-CAM model is focused at the nearest three enemies, and the agent is shooting using LEFTFIRE in the direction of the green bounce craft. This could add extra weight to the idea that the agent does not

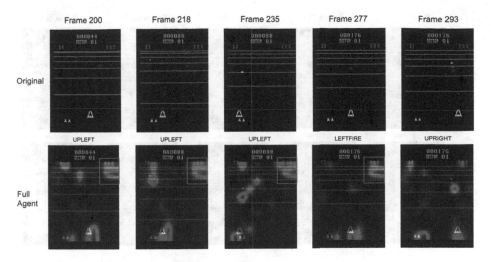

Fig. 5. Manually sampled states from the game BeamRider while not firing torpedoes. Combined with the Full Agent's actions Grad-CAM outputs based on these states. In the 300 frames played it has chosen any UP-variant 219 times, LEFTFIRE 67 and other actions 14 times. Best viewed in high resolution in color. (Color figure online)

know how to use the torpedoes correctly, but perhaps also that the agent might not be able to distinguish one enemy from another; the piece of green debris to the left of the green bounce craft looks quite similar to it.

Seaquest. The agent playing Seaquest has a different problem: it has not learned the long term strategy of surfacing for oxygen. An example of a death due to this is depicted in Fig. 7. The oxygen bar is highlighted by a red box, and it is noticeable that there is no direct or intense activations produced by the rationalization model on the oxygen bar. This could indicate that the agent has never made a correlation between the oxygen bar depleting and the episode ending. A multitude of factors could lead the agent to not learn this such as not having enough temporal knowledge or a lack of exploratory actions. A solution to this could be the use of Fine-Grained Action Repetition which selects a random action and performs this action for a decaying number of times [16].

4.4   Failure Cases of Our Model

Looking at Fig. 7 a prominent activation is depicted in the form of a vertical bar at the top of the state. This vertical bar might seem a bit too ambiguous and even hard to interpret. This type of activation map come in two varieties: activations that highlight only static objects and avoid any non-static objects like agents or enemies and activations that do highlight seemingly at random.

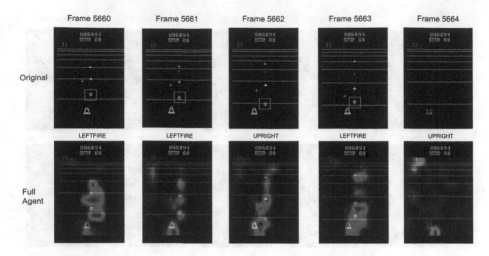

Fig. 6. Agent dies because it is hit by a green bouncecraft, highlighted by the rex box. The green bouncecraft is an enemy that only appears in later sectors of the game, but it looks similar to an enemy which is more easily avoidable and which also appears multiple times in each sector. Best viewed in high resolution in color. (Color figure online)

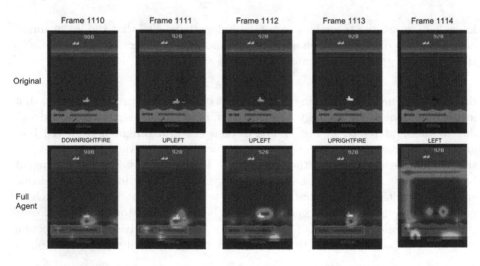

Fig. 7. The agent dies due to lack of oxygen depicted in the red box. Looking at the activation map for the Full Agent, it is noticeable that there are no (direct) Grad-CAM activations on the oxygen bar. This could indicate a lack of understanding of the oxygen mechanism that allows the agent to live longer and get a higher score. Best viewed in high resolution in color. (Color figure online)

The activations that highlight everything except for non-static objects are noticeable in Fig. 8 in the case of Pong and BeamRider. For Pong, the activations are not focused on the ball but on everything except for the ball which could

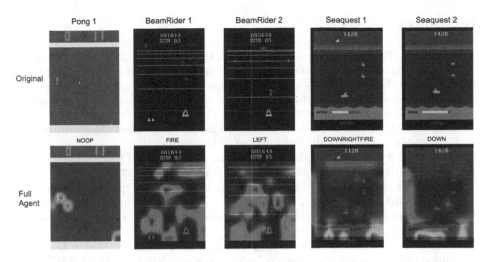

Fig. 8. (Seemingly) ambiguous rationalization outputs. The activations depicted in Pong highlight the agent and its enemy. The activations are also noticeable lightly on the whole field except for the bal itself. When looking more closely to the BeamRider activations, it appears that there are activations surrounding important in the game such as the agent, the lives left and incoming enemies. The Seaquest activations, in contrast to the other games, seem more scattered and not focused on either objects or space between objects. Best viewed in high resolution in color. (Color figure online)

still indicate some pattern for the agent. For BeamRider the activations are highlighting areas directly next to the non-static agent and enemies in the state. This could indicate that the agent is calculating the trajectory of enemies or possible safe locations for it to go to.

The activations that are seemingly at random are depicted in Fig. 8 in the last two Seaquest frames. A possible explanation for this could be that the agent is not provided with enough evidence and is indifferent to taking any action, which is reflected in the ambiguous activation map.

5   Conclusion

In this work, we have presented a post-hoc explanation framework that visually rationalizes the output of a deep reinforcement learning agent. Once the agent has made the decision of which action to take, the model propagates the gradients that lead to that action back to the image. Hence, it is able to visualize the activation map of the action as a heatmap. Our experiments on three Atari 2600 games indicate that the visualizations successfully attend to the regions such as the agent and the obstacle that lead to the action. We argue that such visual rationalizations, i.e. post-hoc explanations, are important to enable communication between users and the agents. Future work will include a quantitative evaluation in the form of a user study or developing an automatic evaluation metric for these kind of visual explanations.

References

1. Andreas, J., Rohrbach, M., Darrell, T., Klein, D.: Neural module networks. In: 2016 IEEE Conference on Computer Vision and Pattern Recognition (CVPR) (2016)
2. Bellemare, M.G., Naddaf, Y., Veness, J., Bowling, M.: The arcade learning environment: an evaluation platform for general agents. CoRR abs/1207.4708 (2012). http://arxiv.org/abs/1207.4708
3. Biran, O., McKeown, K.: Justification narratives for individual classifications. In: Proceedings of the AutoML Workshop at ICML 2014 (2014)
4. Greydanus, S., Koul, A., Dodge, J., Fern, A.: Visualizing and understanding atari agents. CoRR abs/1711.00138 (2017). http://arxiv.org/abs/1711.00138
5. Hausknecht, M., Stone, P.: Deep recurrent Q-learning for partially observable MDPs. CoRR, abs/1507.06527 (2015)
6. Heess, N., et al.: Emergence of locomotion behaviours in rich environments. arXiv preprint arXiv:1707.02286 (2017)
7. Henderson, P., Islam, R., Bachman, P., Pineau, J., Precup, D., Meger, D.: Deep reinforcement learning that matters. arXiv preprint arXiv:1709.06560 (2017)
8. Hendricks, L.A., Akata, Z., Rohrbach, M., Donahue, J., Schiele, B., Darrell, T.: Generating visual explanations. In: Leibe, B., Matas, J., Sebe, N., Welling, M. (eds.) ECCV 2016. LNCS, vol. 9908, pp. 3–19. Springer, Cham (2016). https://doi.org/10.1007/978-3-319-46493-0_1
9. Hessel, M., et al.: Rainbow: combining improvements in deep reinforcement learning. arXiv preprint arXiv:1710.02298 (2017)
10. Mnih, V., et al.: Asynchronous methods for deep reinforcement learning. In: International Conference on Machine Learning, pp. 1928–1937 (2016)
11. Mnih, V., et al.: Human-level control through deep reinforcement learning. Nature 518(7540), 529–533 (2015)
12. Park, D.H., et al.: Multimodal explanations: justifying decisions and pointing to the evidence. In: IEEE CVPR (2018)
13. Schulman, J., Levine, S., Abbeel, P., Jordan, M., Moritz, P.: Trust region policy optimization. In: International Conference on Machine Learning, pp. 1889–1897 (2015)
14. Schulman, J., Wolski, F., Dhariwal, P., Radford, A., Klimov, O.: Proximal policy optimization algorithms. CoRR abs/1707.06347 (2017). http://arxiv.org/abs/1707.06347
15. Selvaraju, R.R., Cogswell, M., Das, A., Vedantam, R., Parikh, D., Batra, D.: Grad-CAM: visual explanations from deep networks via gradient-based localization. In: IEEE ICCV (2017)
16. Sharma, S., Lakshminarayanan, A.S., Ravindran, B.: Learning to repeat: fine grained action repetition for deep reinforcement learning. CoRR abs/1702.06054 (2017). http://arxiv.org/abs/1702.06054
17. Silver, D., et al.: Mastering the game of go with deep neural networks and tree search. Nature 529(7587), 484–489 (2016)
18. Teach, R.L., Shortliffe, E.H.: An analysis of physician attitudes regarding computer-based clinical consultation systems. In: Anderson, J.G., Jay, S.J. (eds.) Use and Impact of Computers in Clinical Medicine. Computers and Medicine, pp. 68–85. Springer, New York (1981). https://doi.org/10.1007/978-1-4613-8674-2_6
19. Van Hasselt, H., Guez, A., Silver, D.: Deep reinforcement learning with double Q-learning. In: AAAI, vol. 16, pp. 2094–2100 (2016)

20. Williams, R.J.: Simple statistical gradient-following algorithms for connectionist reinforcement learning. In: Sutton, R.S. (ed.) Reinforcement Learning. The Springer International Series in Engineering and Computer Science (Knowledge Representation, Learning and Expert Systems), vol. 173, pp. 5–32. Springer, Boston (1992). https://doi.org/10.1007/978-1-4615-3618-5_2
21. Williams, R.J.: Simple statistical gradient-following algorithms for connectionist reinforcement learning. Mach. Learn. **8**(3), 229–256 (1992). https://doi.org/10.1007/BF00992696
22. Xu, B., Wang, N., Chen, T., Li, M.: Empirical evaluation of rectified activations in convolutional network. CoRR abs/1505.00853 (2015). http://arxiv.org/abs/1505.00853
23. Zintgraf, L.M., Cohen, T.S., Adel, T., Welling, M.: Visualizing deep neural network decisions: prediction difference analysis. In: ICLR (2017)

Author Index

Printed in the United States
By Bookmasters